EMS Series in Industrial and Applied Mathematics

Editorial Board:
Alfredo Bermúdez de Castro (Universidade de Santiago de Compostela, Spain)
Lorenz T. Biegler (Carnegie Mellon University, Pittsburgh, USA)
Annalisa Buffa (IMATI, Pavia, Italy)
Maria J. Esteban (CNRS, Université Paris-Dauphine, France)
Matthias Heinkenschloss (Rice University, Houston, USA)
Alexander Martin (Universität Erlangen-Nürnberg, Germany)
Volker Mehrmann (Technische Universität Berlin, Germany)
Stephen B. G. O'Brien (University of Limerick, Ireland)

The *EMS Series in Industrial and Applied Mathematics* publishes high quality advanced texts and monographs in all areas of Industrial and Applied Mathematics. Books include those of a general nature as well as those dealing with the mathematics of specific applications areas and real-world situations. While it is intended to disseminate scholarship of the highest standard, authors are encouraged to make their work as accessible as possible.

Previously published in this series
MATHEON – Mathematics for Key Technologies, *edited by Peter Deuflhard, Martin Grötschel, Dietmar Hömberg, Ulrich Horst, Jürg Kramer, Volker Mehrmann, Konrad Polthier, Frank Schmidt, Christof Schütte, Martin Skutella and Jürgen Sprekels*

Decision Support Systems for Water Supply Systems

Smart Water System to improve the operation of water supply systems by using applied mathematics

Andreas Pirsing
Antonio Morsi

Editors

On behalf of the Editors:

Dr. Andreas Pirsing
c/o Siemens AG
Nonnendammallee 101
13629 Berlin, Germany

andreas.pirsing@siemens.com

2020 Mathematics Subject Classification: 68U07, 76B75, 93C05, 90C11, 90C30, 65L80, 90-08, 90-10, 90-11

Key words: Optimization, simulation, modelling, automation, ICT, decision support system, digital twin, water supply, digitalization, automation, energy management

ISBN 978-3-03719-207-8

The Swiss National Library lists this publication in The Swiss Book, the Swiss national bibliography, and the detailed bibliographic data are available on the Internet at http://www.helveticat.ch.

This work is subject to copyright. All rights are reserved, whether the whole or part of the material is concerned, specifically the rights of translation, reprinting, re-use of illustrations, recitation, broadcasting, reproduction on microfilms or in other ways, and storage in data banks. For any kind of use permission of the copyright owner must be obtained.

© European Mathematical Society 2020

    Contact address:

    European Mathematical Society - EMS - Publishing House GmbH
    TU Berlin Mathematikgebäude
    Room MA266
    Straße des 17. Juni 136
    10623 Berlin
    Germany

    Email: info@ems.press
    Homepage: https://ems.press

Typeset using the authors' TeX files: le-tex publishing services GmbH, Leipzig, Germany
Printing and binding: Beltz Bad Langensalza GmbH, Bad Langensalza, Germany
∞ Printed on acid free paper
9 8 7 6 5 4 3 2 1

Preface

Automation and digitization are becoming more and more important in the water industry. The drivers for using these new technologies are increasing legal requirements, continued high cost pressure and increasing water scarcity in many regions of the world. For this reason, both Germany and the European Union have set up a number of applied and practice-oriented research programs dealing with the development of innovative IT applications and the application of model-based production planning algorithms based on simulation and optimization. The research project EWave, funded by the Federal Ministry of Education and Research (BMBF), pursues the goal of expanding an existing process control system with a decision support system (DSS) (for energy and/or resource efficient operation. Over the past few years, requirements for drinking water supply have become more and more demanding. While the secure supply of high-quality drinking water for the public was the priority during the past decades, rising energy costs and flexible energy tariffs now also require that energy is used efficiently. Water supply companies must therefore rise to the challenge of continuing to prioritize a secure supply of high-quality drinking water while coping with the increasing demand for energy and cost efficiency. Therefore, the operation of drinking water supply systems, such as urban supply networks or long-distance water pipes, is an extremely complex task which can no longer be carried out without the use of modern tools. Today, such supply systems use state-of-the-art technology and are controlled via SCADA systems (Supervisory Control and Data Acquisition) and partly automated function sequences. The operators must ensure a reliable supply of excellent quality drinking water for consumers at reasonable costs. Within the project EWave an innovative energy management system (EMS) has been developed which was trialed at Rheinisch-Westfälische Wasserwerksgesellschaft mbH (RWW), a water supplier with a typical network structure. The new software tool which can be used as decision support system (DSS) is able to devise energy-optimized operating plans for plants for water collection, treatment and distribution within the supply system. Moreover, the fluctuating energy supply from suppliers' own power plants are co-ordinated with the energy supplied by one or more other energy suppliers. The scope for optimization calculation is restricted by technical and operational aspects. Here too, the quality and security of the supply, in particular, must be guaranteed at all times. Water suppliers' electricity consumption is highest for water treatment and water distribution. Therefore, as far as energy optimization is concerned, the focus lays on both specifying the running times and switching times of the network pumps and setting the required water production output to the available water plants. As conditions on the electricity market are changing, shorter contract terms with differing rates and diverse pricing packages can be expected. This means that, on the one hand, new opportunities for optimization open up; on the other hand, operating procedures which have been predominately static so far must become more dynamic. That also implies that knowledge and experience accumulated over decades by the operating personnel is partly devalued and must be further developed. The BMBF project closes this gap by using mathematical optimization methods to determine an operation mode at optimal costs. Parameters developed within the project are used to assess the energy efficiency and profitability of a company. This efficiency assessment allows for a comparison of different plants and waterworks. In addition to energy efficiency, hydraulic efficiency was calculated to assess hydraulic plant components. An overall assessment of the plant is realized by calculating the plant efficiency.

In Part I ICT solutions for water supply systems are introduced in a general manner. The advances in information and communication technology (ICT) have already led to new technological solutions in the water industry. This continues a trend that started long time ago, but in many cases has been limited by several technical restrictions. Chapter 1 describes the ICT basics that are relevant for the water industry and lays the theoretical foundations for understanding the newly developed energy management system. Chapter 2 summarizes the main conditions from the real pilot system to be met during operation and the potentials for improvement to be exploited are translated into various requirements for the development of the new tool. To meet the current requirements for such a DSS, several workshops for water supply companies were conducted. During these workshops, experiences gained with energy management systems were exchanged and further aspects were added to the list of requirements.

Part II deals with theoretical aspects which are essential to develop a decision support system for water supply system. Chapter 3 outlines that knowledge about current and future demands for drinking water and energy is a relevant boundary condition for an optimized operation. Energy demands depend on the hydraulic conditions within the waterworks and the entire distribution network of the relevant pressure zone. Therefore, a water demand forecast tool has been developed. The hydraulic simulation model, described in Chapter 4, is based on a network abstraction that describes network components, such as pipes, storages, valves, or pumps by mathematical equations and connects them with each other in a network graph. The resulting nonlinear, differential-algebraic equation system is then automatically generated and numerically solved with suitable methods. Chapter 5 explains a newly developed integrated optimization approach for decision making and operational support. As a result, the user receives operating schedules on a 15 minute scale. For this purpose, discrete-linear, and continuous-nonlinear mathematical optimization methods are combined. First, a mixed-integer optimization model is solved in order to derive all discrete decisions (primarily pump schedules). Second, these results are used for the discrete optimization variables and subsequently optimizes the continuous variables such as pump speed, valve opening degrees, or water volumes.

Part III describes the application of the new energy management system in a real plant environment at RWW. In Chapter 6 it is shown that network abstraction is necessary in order to ensure that the entire water supply network is considered within the dynamic simulation model. This is realized by combining regional subnetworks to superior pipe and storage elements, thus preserving the large-scale network structure. Chapter 7 deals with the setup of the simulation model. This model consists of two parts: The relevant processes within a waterworks and the distribution network for drinking water. These parts are connected by the drinking water pumps. Calibration is made separately for single network elements like pumps and valves and for the aggregated network. The calibration of e.g. pumps can be done automatically, whereas the calibration of the network is largely a manual process. The achieved accuracy is appropriate for practical application and the further optimization process. Chapter 8 explains the configuration of the real automation system and how existing data is made available to the project partners for model construction, calibration of the models and for test runs of the tool.

Chapter 9 discusses the new ICT architecture of the new DSS that has a modular structure based on existing modules of the SIWA Smart Water Systems by Siemens AG. The calculation starts with an import and pre-processing of current plant data. Subsequently, a simulation run is initiated in order to assess the current status of the plant. In addition, the estimated water demand is calculated. On this basis, the pump schedules are optimized for a preview period of 24 hours and displayed to the user on a graphical user interface (GUI). This process is repeated every 30 minutes. The representation of the results is explained in Chapter 10. Several industry-specific key performance indicators (KPI) are calculated for water works Dorsten-Holsterhausen. The graphical user interface shows the optimized switching points for all pumps in the network in a very easy way. These switching points can be used by the operators to reduce the energy consumption while securing the safety of supply.

Part IV summarizes the pilot application at RWW. Experiences and advantages of the innovative decision support system for water supply systems are discussed. Chapter 11 describes the installation of the tool as pilot application directly on site at the RWW facility in the waterworks Dorsten-Holsterhausen. In a first step, the tool was run in parallel to the real operation, without applying the setpoints on the real plant. After familiarization by the operators, the optimized setpoints were applied to the plant control. In terms of the supply system considered here, optimization yielded an energy savings potential of approximately 10%, based on historical data.

Berlin, *Andreas Pirsing*
Erlangen, *Antonio Morsi*

Acknowledgements

The EWave research and development project (from a German acronym for *water supply energy management system*) is being supported by the German Federal Ministry of Education and Research (BMBF) under project no. 02WER1323A-F as part of the funding measure *Future-oriented Technologies and Concepts for an Energy-efficient and Resource-saving Water Management* (ERWAS). With ERWAS, the BMBF is pursuing the goal of sponsoring applied and practice-based research projects from the field of public water supply and wastewater treatment. The interdisciplinary composition of the project team – comprising industrial partners Siemens AG and Rheinisch-Westfälische Wasserwerksgesellschaft mbH, as well as the scientific partners Friedrich Alexander University of Erlangen-Nuremberg, TU Darmstadt, Bonn-Rhein-Sieg University of Applied Sciences and the University of Mannheim, ensured the practice-oriented application of proven mathematical methods to the water industry. The partners thank BMBF for the financial support. The treatment and distribution of water demands enormous quantities of energy. The aim of the EWave project is to extend an existing process control system with an innovative decision support system. In this way, the developed software tool performs the function of an energy management system.

Contents

Part I. ICT solutions for water supply systems 1

1 Role of ICT in water supply systems: requirements, current status and challenges 3
by Andreas Pirsing, Moritz Allmaras, Roland Rosen, Tim Schenk, and Annelie Sohr

    1.1 Introduction 3
    1.2 Basic concepts of digitalization 4
    1.3 Enabling technologies of digitalization 6
    1.4 Basic IT security technologies 8
    1.5 Decision support systems for water supply systems 11
        1.5.1 Methodological approaches for decision support systems . . . 12
        1.5.2 Practical aspects of decision support systems 13
        1.5.3 Technological components 14
    1.6 Energy aspects 14
    Bibliography 15

2 EWave energy management system 19
by Constantin Blanck, Stefan Fischer, Michael Plath, Moritz Allmaras, Andreas Pirsing, Tim Schenk, and Annelie Sohr

    2.1 Objectives 19
    2.2 Requirements to the water supply 20
        2.2.1 Electricity stock exchange 22
        2.2.2 Atypical grid utilization 23
        2.2.3 Control energy 23
    2.3 Pilot network – Dorsten-Holsterhausen 25
        2.3.1 Water production 25
        2.3.2 Water purification 26
        2.3.3 Water distribution 28
    2.4 Basic concepts of the decision support system 29
        2.4.1 Time aspects 30
        2.4.2 System models 31
        2.4.3 Coupling simulation and optimization 32
        2.4.4 Prognosis 33
        2.4.5 Boundary conditions 33
        2.4.6 Data access 34
    Bibliography 34

Part II. Theoretical aspects 37

3 Demand forecast 39
by Patrick Hausmann
- 3.1 Introduction 39
  - 3.1.1 Necessity of a demand forecast 39
  - 3.1.2 Concept, model approach, and classification 40
- 3.2 Demand forecast 41
  - 3.2.1 Data clustering 41
  - 3.2.2 Model parametrization 42
  - 3.2.3 Artificial increase of time resolution 43
  - 3.2.4 Program flow 45
  - 3.2.5 Results and accuracy 46
- 3.3 Summary 49
- Bibliography 49

4 Hydraulic modeling and energy view 51
by Gerd Steinebach and Oliver Kolb
- 4.1 Introduction 51
- 4.2 Water supply network modeling 52
  - 4.2.1 Component-based network approach 52
  - 4.2.2 Modeling equations 53
  - 4.2.3 Energy view 57
- 4.3 Simulator TWaveSim 57
  - 4.3.1 Coupling conditions within PDAE system 58
  - 4.3.2 Semidiscretization in space and boundary conditions 59
  - 4.3.3 Initial values 61
  - 4.3.4 DAE solver 62
- 4.4 Simulator Anaconda 63
- 4.5 Test application 64
  - 4.5.1 Problem setup 64
  - 4.5.2 Simulation results 67
- 4.6 Conclusion 70
- Bibliography 70

5 Optimization 73
by Björn Geißler, Alexander Martin, Antonio Morsi, Maximilian Walther, Oliver Kolb, Jens Lang, and Lisa Wagner
- 5.1 Introduction 73
- 5.2 Discrete optimization 75
  - 5.2.1 Water supply network model 75
  - 5.2.2 Solution approach 89
  - 5.2.3 Computational results 99
- 5.3 Continuous optimization 100
- Bibliography 102

Part III. Practical aspects 105

6 Network aggregation 107
by Tim Jax
- 6.1 Introduction 107
- 6.2 Theoretical aspects 109
  - 6.2.1 General objectives 109
  - 6.2.2 General concept 111
  - 6.2.3 Assumptions and requirements 113
  - 6.2.4 Layout definition 114
  - 6.2.5 Pipe aggregation 116
  - 6.2.6 Generating artificial tanks 118
  - 6.2.7 Sink realization 120
- 6.3 Generator TWaveGen 120
  - 6.3.1 General aspects 120
  - 6.3.2 Pipe selection 122
  - 6.3.3 Tank generation 124
- 6.4 Conclusion and outlook 125
- Bibliography 127

7 Setup of simulation model and calibration 129
by Gerd Steinebach, David Dreistadt, Patrick Hausmann, and Tim Jax
- 7.1 Introduction 129
- 7.2 Application: model Holsterhausen 130
  - 7.2.1 Modeling waterworks Dorsten-Holsterhausen 130
  - 7.2.2 Modeling pressure zone Holsterhausen 131
- 7.3 Calibration 133
  - 7.3.1 Calibration of characteristic pump curves 133
  - 7.3.2 Calibration of valve coefficients 135
  - 7.3.3 Calibration of network model 137
- 7.4 Model operation and simulation results 143
  - 7.4.1 Input data and program handling 143
  - 7.4.2 Discussion of simulation results 144
- 7.5 Summary 145
- Bibliography 146

8 Field data, automation, instrumentation and communication 147
by Constantin Blanck, Stefan Fischer, and Michael Plath
- 8.1 Data provision, network analysis and acquisition of additionally required measurement and control technology 147
- 8.2 Electricity price forecast 149

9 New ICT architecture 151
by Tim Schenk, Moritz Allmaras, Andreas Pirsing, and Annelie Sohr
- 9.1 Requirements 151
- 9.2 General architecture and model-based approach 153
  - 9.2.1 Overview 153
  - 9.2.2 Process and data flow 155
  - 9.2.3 Structure of the data model 156
  - 9.2.4 The calculation module interface 161
- 9.3 The EWave sequence 165
  - 9.3.1 Overview 165
  - 9.3.2 Simulation state calculation 165
  - 9.3.3 Optimization 167
  - 9.3.4 Simulation forecast 168
  - 9.3.5 Measurement processing 169
  - 9.3.6 Processing of boundary conditions 171
  - 9.3.7 Demand prognosis 175
  - 9.3.8 Cyclic evaluation 177
  - 9.3.9 Evaluation of switch messages 178
  - 9.3.10 Summary 181
- Bibliography 181

10 Water cockpit: dashboards for decision support systems 183
by Michael Plath, Constantin Blanck, Stefan Fischer, Moritz Allmaras, Andreas Pirsing, Tim Schenk, and Annelie Sohr
- 10.1 Energy and process 183
  - 10.1.1 Industry-specific indicators, especially efficiencies 183
  - 10.1.2 Application test of efficiencies 186
- 10.2 User interface technologies 189
  - 10.2.1 User roles 189
  - 10.2.2 System qualities & decisions 190
  - 10.2.3 User interface 190
  - 10.2.4 Communication 196
  - 10.2.5 Technologies 197
- Bibliography 198

Part IV. Outlook 199

11 Field test 201
by Annelie Sohr, Constantin Blanck, Stefan Fischer, Michael Plath, Moritz Allmaras, Tim Schenk, and Andreas Pirsing

    11.1 Implementation and pilot application 201
        11.1.1 Preparation of the pilot application 202
        11.1.2 Test phase I 202
        11.1.3 Test phase II 206
        11.1.4 Drinking water demand forecast test 208
        11.1.5 Conclusion: pilot application 208
    11.2 Comparative calculations 209
        11.2.1 Concept 209
        11.2.2 Verification 211
    11.3 Conclusion & outlook 215
    Bibliography 216

Acronyms 217

Symbols and Parameters 219

Glossary 221

List of Contributors 223

Index 225

Part I

ICT solutions for water supply systems

The operators of drinking water networks face the contrasting challenges of supply reliability and cost efficiency. On the one hand, customers expect a reliable supply with high-quality drinking water while, on the other hand, any price increases are difficult to implement. The EWave research project, sponsored by the German Federal Ministry of Education and Research (BMBF), pursues the aim of extending an existing process control system with a partially automated decision support system to achieve energy-efficient operational management. The power consumption of a water supply company is essentially determined by the functions of water procurement, purification and distribution. Optimization of the operational energy use should therefore enable the following results in particular to be calculated, e.g. runtimes or switching times of the grid pumps (pump timetables), reservoir plans and / or distribution of the required production volume to the available water recovery and purification plants (production plans). A key sub-function in the context of the change to renewable energies is the coordination of the volatile supply of energy from one's own sources with the energy procured from one or more energy supply companies. Due to the interlinked network structures and technological restrictions of the network components, an extremely complex system with networked interdependencies is created from the viewpoint of the operating personnel. The operational management today is performed almost exclusively by experienced personnel and demands very precise knowledge of the supply network, of the in-house energy generation, and of the power supply contracts signed with one or more energy supply companies. It is to be expected that the foreseeable changes in the energy market due to the energy revolution will lead in future to shorter contract periods with a greater variety of differentiated tariffs. The positive aspect of this development is that it will give rise to new opportunities for optimization. Conversely, however, there is a danger for the water supply companies that the knowledge and experience of operating personnel gained over many years will be devalued to some extent. The new software tool developed within the EWave project offers water supply companies support in managing this structural change. A general list of requirements of water supply serves as a basis for the development of a simulation-based assistance system.

Chapter 1

Role of ICT in water supply systems: requirements, current status and challenges

Andreas Pirsing, Moritz Allmaras, Roland Rosen, Tim Schenk, and Annelie Sohr

Abstract. The advances in information and communication technology (ICT) have already led to new technological solutions in the water industry. This continues a trend that started long time ago, but in many cases has been limited by several technical restrictions. Digitalization is also changing economic activities in many ways: workflows and forms of organization are being transformed to the emergence of new business models. This chapter describes the ICT basics that are relevant for the water industry and lays the theoretical foundations for understanding the newly developed decision support system EWave.

1.1 Introduction

The digital economy, like many other industries, is facing fundamental structural change. All industrial sectors are affected like, manufacturing and process industry, medical technology, aviation technology, automotive engineering and logistics. The experience from recent years shows that this structural change is unstoppable and new business models in the internet age are disruptive and change markets completely as seen in online media. The different industries differ only in terms of the beginning and the speed with which sustainable changes occur.

The research project EWave addresses key aspects of digitalization. Before presenting the industry-specific interpretation of digitalization for the operation of water supply systems, the general concept of digitalization is first explained in order to provide a theoretical basis for the later explanations.

Digitalization in Germany is often equated with the term "Industry 4.0", which goes back to the project of the same name in the action plan for the High-Tech Strategy 2020 [28] of the Federal Government. The term was coined at the Hanover Fair 2011 when handing over the recommendations for action of the Working Group on Industry 4.0. Two years later, the final report of the working group [46, 45] was handed over to the Federal Government. In order to coordinate further work, the Industry 4.0 platform launched by the three industry associations Bitkom, VDMA (Verband Deutscher Maschinen- und Anlagenbau) and ZVEI (Zentralverband Elektrotechnik-

und Elektronikindustrie) was launched in 2013. Two years later, the platform for representatives of companies, associations, trade unions, science and politics also opened [37, 47]. At the Hannover Fair 2016, the platform presented its first progress report in the extended set-up [14].

The Industry 4.0 initiative has met with broad international interest, taking advantage of current technological trends [10, 49, 23]. Similar work is being pursued in the USA under the terms "Advanced Manufacturing" [3, 39] and Industrial Internet Consortium [1] as well as in China under the term "Made in China 2025" [51].

And, the technological development continues and leads to new cooperation opportunities as the Industrie 4.0 cooperation between China and Germany [13].

The term "Industry 4.0" coined by the authors' recommendation was originally only understood as a synonym for the use of new digital possibilities for production. In the meantime, many authors see the need to broaden the concept and, in addition to the technical aspects, to consider the associated changes in the business models used [18] as well as working conditions [43]. This quickly makes it clear that Industry 4.0 will lead to the digital transformation of entire branches and industries [29, 7]. For this reason, the authors favor the term "digitalization" and thus follow the understanding of the Steering Committee of the Plattform Industrie 4.0 [47]: "The term Industrie 4.0 stands for the fourth industrial revolution, a new stage in the organization and control of the entire value chain via the life cycle of products. This cycle is geared to the increasingly individualized customer requirements and extends from the idea, the order to the development and production, the delivery of a product to the end customer to recycling, [...]. It is based on the availability of all relevant information in real time by connecting all the entities involved in the creation of value as well as the ability to derive the data from the optimal flow of value at any time."

This term is intended to express that the increased use of Information and Communication Technology (ICT) in the automation of production processes leads to a fourth stage of the industrial revolution. The key point of digitalization is that Internet-based technologies not only enable the connection of production systems, IT systems and humans, but also create the conditions for innovative production organizations and business models.

Digitalization is also receiving so much attention in the public debate because it is becoming increasingly clear that under the ever-increasing influence of ICT within a short time, a profound change will take place in almost all areas of the economy and society. In many cases, these new technologies have the potential to completely displace existing products, services or solutions [2, 5, 25, 32].

1.2 Basic concepts of digitalization

An essential cornerstone of digitalization is Internet-based information and communication systems, which enable the comprehensive networking of all involved entities to value chains or networks [34, 41, 42]. The "Industry 4.0" platform defines this: "By connecting people, objects and systems, dynamic, real-time-optimized and self-organizing, cross-company value chains are created that can be optimized according to different criteria such as costs, availability and resource consumption."

The creation of value chains belongs to the three main concepts of digitalization:

- Horizontal integration
  Under this, the combination of different IT systems for the establishment of integrated value-added processes takes place. Modern ICT enables this within a company as well as across company boundaries.

- Vertical integration
  This means the merger of different IT systems on the hierarchy levels of the automation pyramid into an integrated solution. In the field of process automation of water supply systems this includes the field level, control level, process control level, operations level and enterprise level.

- Integrated engineering over the entire life cycle
  This means the aggregation of all information that accumulates along the life cycle of a project. This creates a digital twin of the water supply system that can be used to support real-world decision-making.

The use of these concepts not only creates the conditions for completely new production organizations and business models, but also leads to a fundamental change in the organization of work.

Horizontal integration. Horizontal integration means that the interaction of cyber-physical systems does not end at an enterprise boundary. Instead, the future project Industry 4.0 supports the development of value chains beyond company boundaries [26]. Only the creation of these value chains makes it possible for companies to react flexibly to changing operating conditions.

Vertical integration. Vertical integration is the seamless integration of systems from the field level to the enterprise level, which describes the interconnectivity of intelligent production systems and IT systems in a company. Of particular importance is the integration of Manufacturing Execution Systems (MES) and Enterprise Resource Planning Systems (ERP). As a result, the data from the different business units are linked together and data is made available electronically at all company levels. These aspects are often described by the term Internet of Things (IoT) and related to the rapid development of cloud platforms and solutions. Two specific aspects of vertical integration are that they allow IT systems to be set up for real-time control and / or optimization to adapt operations management to current conditions.

The IoT approach leads to a significant increase in the sensors used in a plant, which record the current state of the process and the production systems online. With this comprehensive information, production control procedures can be set up for on-line control and optimization. In addition, the data can be used to derive forecasts about the future behavior of processes and production systems and to initiate corresponding actions [17, 21].

Integrated engineering over the entire life cycle. Integrated engineering is one of the cornerstones of the future project Industry 4.0, as a benefit can only be generated with a consistent flow of information between the value-adding processes [15]. Integrated engineering covers two essential aspects [36].

- Digital twin to integrate real and virtual worlds
  As part of the future project Industry 4.0, cyber-physical systems will be equipped with a digital image in the form of mathematical simulation models that ideally represent the real world and are linked to it. In this way, decision-making processes of the real world can be supported [9].

- Model-based systems engineering
  Relevant data and models are not only used in-house but also exchanged and enriched across company boundaries. On the one hand, this makes it possible to validate plant and operating concepts with the aid of these models in early life cycle phases, on the other hand, findings from later phases (e.g., from the operating phase) are integrated into the models and taken into account in future decision-making [8].

1.3 Enabling technologies of digitalization

The vision formulated by Platform Industry 4.0 states that in the course of digitalization, machines, systems or sensors will in future, communicate with each other and exchange information [3]. As a result, water supply companies can not only make their operations much more efficient, but they can also be much more flexible for external conditions, such as react to a volatile energy supply.

Digitalization uses the new possibilities of innovative technologies of information and communication technique for this transformation and the associated value-added potentials, which are therefore also called core-technologies of Industry 4.0 [6]:

- Big data and data analytics
- Cyber-physical systems (CPS)
- Internet of things
- Internet of services

The use of these core technologies makes it possible for machines and systems to network together and to set up a comprehensive monitoring and evaluation of the underlying processes with the resulting data. These core technologies are already widely used today independently of each other. Industry 4.0 brings these technologies together and supports their widespread use [4].

Big data and data analytics. The extensive implementation of additional sensors and their networking results in large amounts of data that can be used to control and optimize plants [6, 30, 50]. For example, analysis of the historical warnings and alert messages of aggregates, e.g. motors, valves, etc., to derive predictions about the expected behavior of these plants and to perform preventive maintenance actions leading to increased availability [31].

On the one hand, the use of this data opens up new areas of application, but on the other hand it also faces companies with great challenges [18]. The data is generated in large quantities, great diversity and high speed, so that the existing IT systems often reach their limits.

Big data is a new field of knowledge dealing with the processing of large amounts of data [12]. Data analytics is a subset of big data that focuses on gaining insights from data to make better and faster decisions [16]. By using data analytics, companies can understand, predict and optimize their business processes. The consulting firm Gartner describes data analytics as one of the core technologies for Industry 4.0, since analytics is the basis for improved business processes and new business models [40].

Data mining is also a subset of big data and provides methods for extracting knowledge from large amounts of data [24]. Another newly (re-)discovered approach is artificial intelligence (AI), which tries to automate intelligent behavior and uses machine learning approaches like neural networks. Data and big data are a source to derive and to learn the behavior.

In addition to the new possibilities, the processing of large amounts of data also poses completely new challenges for the industry, because the data is generated in large quantities and with a very high temporal resolution [18]. As a result, the global database has been growing rapidly for several years: in 2005, it was about 130 exabytes and increased to about 462 exabytes by 2012. By 2020, data volume is expected to increase further to 14,996 exabytes (around 15 trillion gigabytes). In order to correctly evaluate and use these gigantic amounts of data, you first have to be able to process them. However, the currently available industrial IT systems already often reach their capacity limits.

Cyber-physical systems. Cyber-physical systems (CPS) are a central component of the future project: Industry 4.0. CPSs are embedded systems that have the ability to connect to other CPSs on its own. Embedded systems result from a combination of a technical system and an IT system. In addition, CPSs have sensors and actuators to communicate and exchange data with each other. With these skills, they are particularly suitable for monitoring and controlling physical processes [33].

In summary, the merging of the physical-analog world and the information technological or digital world [11] is the eponymous feature of cyber-physical systems. In contrast to classical aggregates, CPSs not only have a physical representation but also a digital one. The digital representation can also be understood as a virtual image that describes the characteristics and the behavior in the form of data and models [33, 11].

However, the sole merger of analog and discrete worlds into CPS is not enough to exploit the full potential of the future project Industry 4.0. Only networking with other CPSs via local or global networks opens up new possibilities in a wide variety of application areas. When CPSs are used in the industrial environment, cyber-physical production systems (CPPS) are also often talked about. The key features of these CPPS are their high degree of adaptability and flexibility: By replacing individual components or entire plant components, Industry 4.0 can react very flexibly to changing conditions of production, whereby the exchange through self-organization in the form of a "Plug & Produce" concept is strong and is simplified. In addition, CPPS enables automatic optimization of operations management and supports diagnostic functions.

The use of cyber-physical systems requires up-to-date information, methods and algorithms as well as sufficient memory and computing capacity for all involved components [6]. As CPS places significant demands on the information technology infrastructure, Internet-based cloud solutions are increasingly becoming established platforms.

As a result, the advent of cloud solutions in the industrial environment means that the use of CPSs leads to a disintegration of the classical automation hierarchy. The use of CPPS causes a shift of functions of classical production systems to decentralized units. This leads to an increasing resolution of the automation pyramid and creates a network of CPSs [23].

Internet of things. Plants and aggregates evolve into cyber-physical (production) systems that sense their own state and that of their environment through sensors and communicate with other systems through the Internet of Things. The resulting data forms the basis for data-driven business processes and new business models.

Through vertical and horizontal integration, the future project Industry 4.0 supports cross-value-added communication and cooperation between a wide variety of components. Since very different production and IT systems are involved in this communication, the standardization of communication and data exchange is of central importance. For this purpose, suitable interaction mechanisms and communication models are needed, which i.a., define the use of a common semantic model for information and connectivity to any network nodes.

1.4 Basic IT security technologies

For a long time, water supply systems were not considered the target of cyberattacks. But the introduction of Ethernet, wireless LAN, telecontrol and web-based machine services such as remote monitoring, remote administration, remote diagnostics and remote maintenance not only leads to technical improvements but is also associated with a high dependence on the OT systems (Operational Technology, e.g. automation) and IT (Information Technology) systems and the associated risks [19]. Not only are the systems themselves exposed to cyber-security threats, but also the provision of the supply task. It should be critically evaluated that the damage potentials multiply, since with the same threat situation as for other companies a particularly high damage potential exists. For this reason, the greatest attention is to be paid to industrial security or plant security when implementing digitalization projects in water supply [20].

In addition, OT systems that differ significantly from traditional office or data center IT are frequently used in the water supply environment. Industrial control systems in combination with industry-specific IT systems are customary. Protecting these systems against cyber-threats requires special treatment that must address the potential for damage while satisfying the specific operational requirements for availability and reliability.

The advantage of this approach lies in the transparent and fast decision basis for incidents, errors and trends. However, there is also a significant drawback: the more open, compatible and standardized the systems become in the facilities, the more vulnerable they are to cyberattacks. Therefore, companies have to re-evaluate existing security concepts. However, security concepts from classic IT cannot be directly transferred to automation components, as significantly shorter response times must be guaranteed in plant operation than in the office environment. In addition, the systems run around the clock, which gives the maintenance team limited opportunities to keep the security measures up to date.

The operators of critical infrastructure systems (KRITIS) have to worry about the value of security for them and if they have state of the art defense technology in place to protect their systems against known and unknown threats [22, 48]. It is important to realize that security cannot be bought, it has to be created. It is not enough to provide security once and then run the equipment; rather safety must be constantly maintained. This is a process. Especially highly specialized attacks on individual complex systems must be different so far encountered. Companies need to reassess existing security concepts and procedures. Security solutions for process engineering systems require a consistent security concept, strict organizational rules as well as continuous monitoring of all communication facilities and clearly defined access.

The security concepts to be implemented are based on the requirements of the respective water supply company. Proven concepts and well-trained personnel already exist in the office IT environment. However, with regard to industrial networks, there is a lack of such sensitized personnel. Using vulnerabilities to gain access to the network is often not very hard for attackers and unauthorized persons. An increasing awareness of security risks is needed. Therefore, constant education and training of employees is essential. Based on the ISO27001 and IEC62443 and the included defense-in-depth model, automation technology was developed to provide comprehensive security concepts [44].

The basic idea of these concepts is to define a procedure that achieves an optimum security level considering the best known security standards for industrial control systems. The concept is based on a multi-pillar approach. First of all, the development of a comprehensive security management program for process control is in the foreground. In this program the organizational and technical basics are laid. The second step involves the identification of all affected plants (of all plants), the analysis of the risk and the derivation of protective measures that minimize identified risks [27, 38].

Hardware protection measures. The first safeguard for the individual automation components is to limit the network communication to the required level. This preserves the benefits of end-to-end communication while reducing risk. On the basis of the so-called cell protection concept [1] (see also IEC62443), automation components can be grouped logically or based on their communication and separated from the rest of the network via firewalls, which can offer a DMZ (demilitarized zone) and other security components. In addition, modern automation components have already been designed to be so robust and equipped with defined communication mechanisms that their control functions, even in extreme situations such as denial-of-service attacks or targeted attacks on individual communication services, are processed as usual. This applies both to the lower communication layers (layers 2 to 4) with their Ethernet and IP protocols and to the application protocols above them (layers 5 to 7).

Software-side access protection. While secure features such as network robustness are automatically available, protection mechanisms such as access protection via VPN (Virtual Private Network) and the firewall must be additionally configured. This initially requires a certain amount of effort, but later avoids disruptions and time-consuming troubleshooting of security-related incidents. The implementation of these measures is supported by the leading manufacturers of automation systems with special security functions. So that unauthorized persons do not make changes to the PLC program or the configuration via the engineering software, the automation components are provided with access protection mechanisms. Different levels of protection apply: On the one hand, functional areas can be protected by assigning passwords. On the other hand, new functions or corrections can be quickly integrated via firmware updates. To detect firmware tampering or sabotage, digital signatures are supported. This allows a target device to check the firmware file to install and accept only authentic updates.

Continuous monitoring. The third step of the security concept also restores the relationship to the security management process. Security is not a one-time project, but one that has to adapt to changing threats. Therefore, it is necessary to establish a continuous monitoring to identify risks and react proactively. The security-related information is routed to a central cyber security operation center that can detect simple and complex attacks. This is done through event correlation software and experienced analysts, who evaluate events and then trigger activities accordingly when an event is conspicuous.

Safety in the waterworks. By putting the concept described above into practice, one can operate water supply systems without sacrificing comfort and safety. Efficient communication mechanisms allow the integration of hitherto self-sufficient control systems into a network that enables parallel, plant-wide operation and monitoring of several self-sufficient systems. A good compromise between security and openness is provided by firewalls, which form a perimeter network, and restrict access via firmly defined ports, applications and services. Access from the internet or the corporate intranet is only released to dedicated computers via the perimeter network and also known and unknown threats like virus, malware or even zero day exploits can be detected and prevented.

In the perimeter network, for example, there is a terminal server with a virus scanner installed, which allows access to other components. Other computers could be a quarantine station, which, when transferring data to the system, checks whether the data is free of viruses before they are transferred to the network. Only users who have a valid certificate and trust the company will be granted access to the network. If the users are uniquely identified, they only have access to the systems released for them and may only move in accordance with the release rules (authorization). This prevents further access to other plant components or the company intranet. All accesses that are made from the outside to the water management system can be encrypted by VPN and logged for inspection.

The path to a secure solution can be supported by a risk and vulnerability assessment (in accordance with IEC62243) that identifies the vulnerabilities of a plant and links them to threat scenarios. On this basis, the risk is determined and derived are the measures to reduce the risk to an acceptable level [35].

1.5 Decision support systems for water supply systems

For a long time, various approaches have been known under the name decision support system (DSS), which strive for the best possible utilization of the existing storage volume in the pipe network. The following operational objectives are pursued:

- optimize operating and energy costs, for example, to avoid pumping at inappropriate times
- protect all plants and assets.

So far, the use of DSS in water supply systems has been discussed mainly by science and research. In practice, however, the potential of improved control strategies is still not being used or only insufficiently exploited. The reasons for this lie, on the one hand, in the high investment costs and on the other hand, in the still existing skepticism towards partially automated control processes.

In supervised control rooms, one usually relies on the operating personnel, whose experience in most cases ensures a correct assessment of the operating status and the management interventions based thereon. However, this form of monitored control also has serious disadvantages: On the one hand, the operating experience of the operators is only slowly formed. On the other hand, it is also very difficult to pass on this operator know-how. However, the biggest challenge arises from the fact that operating personnel are faced with a large number of decisions in the shortest possible time in the event of unusual operating conditions and, as a rule, have no way of weighing up the advantages and disadvantages of different operating variants.

An optimal operating variant, in addition to minimizing the energy requirement for operational management, also takes into account further operational criteria while maintaining security of supply, and can only be found with the help of optimization calculation methods. The production control of the water supply company is formulated as a mathematical optimization problem and solved with best practices.

This procedure offers the advantage that the best possible solution is always found, regardless of the complexity of the network considered and the amount of operator know-how introduced. However, optimization approaches to users and authorities are skeptical, because the calculation results cannot be easily understood.

1.5.1 Methodological approaches for decision support systems

Local controls are characterized in the reservoir levels or flows are kept at a fixed setpoint. In order to minimize the expense of installation, self-regulating control devices are often used. Systems with local control, in contrast to uninfluenced systems, are characterized by the fact that the reservoir level is always maintained. Due to the low implementation costs, local controls always represent a favorable mode of operation when only one reservoir is present in the considered water supply system.

In contrast, central controllers take into account the operating states of all network elements and coordinate their activation in the entire service area via the variation of the setpoints. The calculation of these setpoints, which is transferred to the subordinate local controllers, can be carried out by means of mathematical optimization methods, linear multivariable controllers, rule-based methods (fuzzy controllers, expert systems or decision matrices) or by free programming.

Mathematical optimization methods. Control strategies based on mathematical optimization methods are characterized on the one hand by their performance and on the other hand by their high degree of flexibility. The modular design, in conjunction with block libraries for recurrent process components, allows a high degree of reuse. The adaptation to new water supply systems is done solely by configuring the optimization problem by combining the elements of the libraries, which contains mathematical models of the dynamic behavior of each process component. The adaptation effort for customizing of specific network layouts and parameters is low.

The control task is regarded as a dynamic optimization problem that is processed time-discretely: The optimization results calculated per time step are transferred to the subsystems and act there as setpoints for the local controls over the entire time interval. Since the optimization methods also always include results for subsequent time steps, it is also possible to transfer these values to the underlying subsystems. Thus, setpoints are also available there if the transmission paths are disturbed. Overall, this results in a rolling schedule.

Linear multivariable controllers. Multi-variable controllers calculate several control signals from several measured values. Decisive for the success of multivariable controllers is finding a suitable control law. Due to the linear operation of the controller and the ability of multivariable controllers to control optimization based on optimization is difficult to grasp given the restrictions. However, the simple, manageable structure and the relatively low demands on the performance of the control computer are advantageous.

Rule-based procedures. Rule-based methods, such as Fuzzy controllers or expert systems, try to directly represent the experience of the operating staff and are based on the evaluation of meaningful rules. However, the identification of a suitable rule set cannot be standardized and thus largely left to the skill and prior knowledge of the designer. In the case of complex networks and / or later expansions, this approach, which initially leads easily and quickly to success, quickly reaches its limits. In particular, when taking into account failures of individual control units, be it through defects, maintenance or conversion measures, the number of necessary rules quickly reaches an unmanageable level. A reuse and further use of rule-based procedures is also considered to be low.

Free programming. In the simplest case, it is also possible to implement control strategies without using any methodology. The decision rules are then implemented in the form of "if-then-else" statements directly in the program code. In complex network structures, these programs are not only very confusing, they are therefore difficult to maintain or expand. Reusability becomes almost impossible. Nonetheless, even such heuristic control strategies can lead to a marked improvement in operational management.

1.5.2 Practical aspects of decision support systems

The success of a DSS in business practice not only depends on the performance of the calculation methods used, but is significantly influenced by the user-friendliness and flexibility of the selected software solution.

Graphical user interface. In addition to the handling of technological components, the user interface plays an important role in the required simple and functional operation. A graphical user interface for operation and parameterization enables the user to record all relevant information for the optimization of the operational management as quickly as possible and also to recognize correlations.

In addition to an overall overview of the water supply system, the user interface also has an illustration of the individual network elements, e.g., pumps, valves, wells, etc. The display is very user-friendly and provides the operator in a simple and fast way an overall view of the information relevant to the operation management. In addition to the overall display, the calculated setpoints can be displayed in detail windows.

A special feature is the full integration into the control system. This gives the user a system that, in addition to the classic tasks of control and automation technology, also has intelligent functions to support the operation management. This considerably reduces the training required by the user.

Integration of optimization in the plant control. The calculated optimum values are based on the current network status and the consumption forecasts. These data are at least partially available in the control system as they are determined or predicted using measurements. So an interface from the control system to the optimization which passes on these values to the optimization is necessary.

There are different levels of integration of optimization into the plant control. In the simplest case, the optimization can be used offline or with reading process connection, i.e., the calculated results serve as a suggestion for the operator who can manually enforce the specifications, or they are used for comparative calculations e.g. for plant design planning. The values can also be transferred to a simulation of the real plant, which shows the resulting network states. This can be achieved, for example, by connecting the optimization to a suitable simulation tool, which allows a quick visualization of the calculated management strategy. Plant control decisions can then be made by the personnel in the control room based on the simulated impacts.

In a simulation, the effects of changes in the subgoals can be tested as well as changes in parameters in the water supply system (capacity increases, power supply failure) and various load events.

In the full configuration, an interface from the optimization to the control technology is conceivable, i.e., the results of the optimization can be used directly to control the plant and act as a target flow rate, compliance with which must be ensured by the local control technology. If they are also passed on to a simulation, a simple check by the staff is possible.

1.5.3 Technological components

Due to the practical requirement that the modeling of the optimization problem should be applicable to different water supply systems, a component-oriented approach is used. The technological components are grouped together in a block library and can then be connected, simulating the real network. A graphical plant editor is used for this purpose.

1.6 Energy aspects

For the extraction, processing and distribution of drinking water, large amounts of energy are needed and at the same time CO_2 is released. Innovative ICT solutions in the field of water treatment must therefore reconcile treatment technologies, energy consumption and environmental compatibility. The challenge is to improve the process and energy efficiency of water companies by both developing new technologies and optimizing existing ones.

Drinking water is usually obtained from groundwater resources, springs and surface waters. The supply is divided into the following process steps:

- water extraction from sources, groundwater or bodies of water (rivers and lakes)
- water treatment, in waterworks

- water transport to elevated tanks or directly into the network
- water distribution to the consumption points

The operation of the plants required for the water supply requires energy. The bulk of it is used to transport water to treatment stations, storage tanks or consumers. The greater the height difference between the consumption points and the places of extraction, the higher the energy requirement for the water supply.

The water industry is one of the largest municipal energy consumers in Germany: all systems for public water supply and wastewater treatment have an energy requirement of 6.6 TWh of electricity per year. With this amount of energy, about 1.6 million four-person households could be supplied with energy. The drinking water supply accounts for about 2.4 TWh per year.

Water management systems can create new opportunities to optimize water management plants in terms of their energy balance and the associated use of resources. In addition, such ICT systems enable the integration of water management systems into the water and energy infrastructure of the future. The increased use of renewable energy, often associated with rapidly changing energy availability and short-term fluctuating energy prices will accelerate the use of water management systems with decision making functions for energy- and cost-optimized operation of water supply systems.

Bibliography

[1] http://www.iiconsortium.org. Online, accessed on 21.06.2018.

[2] E. Abele, R. Anderl, J. Metternich, and A. Wank, Effiziente Fabrik 4.0. *Zeitschrift für wirtschaftlichen Fabrikbetrieb* (2015), 150.

[3] *Advanced Manufacturing*, A Snapshot of Priority Technology Areas Across the Federal Government, Subcommittee for Advanced Manufacturing of the National Science and Technology Council, 2016.

[4] W. Bauer, O. Herkommer, and S. Schlund, Die Digitalisierung der Wertschöpfung kommt in deutschen Unternehmen an. *Zeitschrift für wirtschaftlichen Fabrikbetrieb* (2015), 68.

[5] T. Bauernhansl, Revolution geht weiter. *Werkstattstechnik* (2014), 105.

[6] T. Bauernhansl, M. Hompel, M., and B. Vogel-Heuser, *Industrie 4.0 in Produktion, Automatisierung und Logistik*, Springer Vieweg, Wiesbaden, 2014.

[7] R. Berger, *Die Digitale Transformation der Industrie* – Eine europäische Studie von Roland Berger Strategy Consultants im Auftrag des BDI, München, Berlin, 2015.

[8] S. Boschert, C. Heinrich, and R. Rosen, Next Generation Digital Twin. In *Proceedings of TMCE 2018*, 7–11 May, 2018, (I. Horvath, J. P. Suarez Rivero, and P. M. Hernandez Castellano, eds.), Las Palmas de Gran Canaria, Spain, 2018.

[9] S. Boschert and R. Rosen, Digital Twin – The Simulation Aspect. In *Mechatronic Futures Challenges and Solutions for Mechatronic Systems and their Designers*, Springer, 2016.

[10] M. Brettel, N. Friederichsen, M. Keller, and M. Rosenberg, How Virtualization, Decentralization and Network Building Change the Manufacturing Landscape: An Industry 4.0 Perspective. *International Journal of Mechanical, Aerospace, Industrial and Mechatronics Engineering* (2014), 37.

[11] M. Broy, *Cyber-physical Systems Innovation durch Software intensive eingebettete Systeme*, Springer, Berlin, Heidelberg, 2010.

[12] T. H. Davenport, Competing on Analytics. *Harvard Business Review* (2006), 84.

[13] *Deutschland und China: Neue Schritte in der Industrie 4.0 Kooperation*. https://www.plattform-i40.de/I40/Navigation/DE/In-der-Praxis/Internationales/Deutsch-Chinesische-Kooperation/Deutsch-Chinesische-Kooperation.html Online, accessed on 21.06.2018.

[14] *Digitalisierung der Industrie – Die Plattform Industrie 4.0*, 2016.

[15] *Durchgängiges Engineering in Industrie 4.0-Wertschöpfungsketten*, VDI/VDE Statusreport, 2016.

[16] D. Dursun, *Real- Word Data Mining*, Pearson Education, New Jersey, 2014.

[17] M. Elsweier, P. Nyhuis, and R. Nickel, Assistenzsystem zur Diagnose in der Produktionslogistik. *Zeitschrift für wirtschaftlichen Fabrikbetrieb* (2010), 562.

[18] V. Emmrich, M. Döbele, T. Bauernhansl, D. Paulus-Rohmer, A. Schatz, and M. Weskamp, *Geschäftsmodell-Innovation durch Industrie 4.0*, Dr. Wieselhuber & Partner GmbH und Fraunhofer IPA, 2015.

[19] N. Engelhardt, *Cybersicherheit-Wasserwirtschaft auf dem Prüfstand*, BDEW-Seminar, Hamburg, 2014.

[20] N. Engelhardt, *IT-Sicherheit in der Wasserwirtschaft*, 7. DWA-Wirtschaftstage, Bonn, 2015.

[21] P. Engelhardt, M. Weidner, W. Einsank, and G. Reinhart, Situationsbasierte Steuerung modularisierter Produktionsabläufe., *Zeitschrift für wirtschaftlichen Fabrikbetrieb* (2013), 987.

[22] *Gesetz zur Erhöhug der Sicherheit informationstechnischer Systeme – IT-Sicherheitsgesetz*, Bundesministerium des Innern (BMI), 2015.

[23] N. Gronau, Der Einfluss von Cyber-Physical Systems auf die Gestaltung von Produktionssystemen. *Industrie Management* (2015), 16.

[24] J. Han, M. Kamber, and J. Pei, *Data Mining – Concepts and Techniques*, Morgan Kaufmann, 2011.

[25] O. Herkommer and K. Hieble, Ist Industrie 4.0 die nächste Revolution in der Fertigung? *Industrie Management* (2014), 42.

[26] O. Herkommer, M. Kauffmann, and K. Hieble, Erfolg der Produktion von morgen. *Zeitschrift für wirtschaftlichen Fabrikbetrieb* (2014), 153.

[27] *ICSSecurity-Kompendium*, Bundesamt für Sicherheit in der Informationstechnik (BSI), 2017.

[28] *Ideen. Innovation. Wachstum, Hightech-Strategie 2020 für Deutschland*, Bundesministerium für Bildung und Forschung (BMBF), 2010.

[29] V. Koch, R. Geissbauer, S. Kuge, and S. Schrauf, *Chancen und Herausforderungen der vierten industriellen Revolution*, PwC Strategy &, 2014.

[30] W. Kong, L. Li, F. Qiao, and Q. Wu, *Network Manufacturing in the Big Data Environment*, IEEE International Conference on System Science and Engineering, 2014, 13.

[31] J. Krumeich, D. Werth, P. Loos, and S. Jacobi, *Big Data Analytics for Predictive Manufacturing Control – A Case Study from Process Industry*, IEEE International Conference on Big Data, 2014, 530.

[32] T. Kuprat, J. Mayer, and P. Nyhuis, Aufgaben der Produktionsplanung im Kontext von Industrie 4.0. *Industrie Management* (2015), 11.

[33] E. A. Lee, *Cyber Physical Systems: Design Challenges*, IEEE International Symposium on Object and Component-Oriented Real-Time Distributed Computing, 2008, 363.

[34] J. Lentes, H. Eckstein, and N. Zimmermann, amePLM – eine Plattform zur Informationsbereitstellung in der Produkt- und Produktionsentstehung. *Werkstattstechnik* (2014), 146.

[35] S. Lenz, Vulnerabilität Kritischer Infrastrukturen. *Forschung im Bevölkerungsschutz*, Bd. 4, Bundesamt für Bevölkerungsschutz und Katrastrophenhilfe, Bonn, 2009.

[36] L. Libuda, G. Gutermuth, and S. Heiss, Arbeitsabläufe in der Anlagenplanung optimieren. *atp edition-Automatisierungstechnische Praxis* (2011), 40.

[37] *Memorandum der Plattform Industrie 4.0*, 2015.

[38] *Process Control System Security Guidance for the Water Sector*, American Water Works Association (AWWA), Washington D.C., 2014.

[39] *Accelerating U.S. Advanced Manufacturing*, Report to the President Accelerating U.S. Advanced Manufacturing, Washington D.C., 2014.

[40] B. Schmarzo, *Big Data: Understanding How Data Powers Big Business*, John Wiley, Indianapolis, 2013.

[41] R. Schmitt and P. Beaujean, Selbstoptimierende Produktionssysteme. *Zeitschrift für wirtschaftlichen Fabrikbetrieb* (2007), 520.

[42] A. Schuldt and J. Gehrke, Software-Plattformen für die kommende Industrie 4.0. *Industrie Management* (2013), 29.

[43] D. Spath, O. Ganschar, S. Gerlach, M. Hämmerle, T. Krause, and S. Schlund, *Produktionsarbeit der Zukunft*, Fraunhofer IAO, 2013.

[44] L. Terhart and K. Wagner, IT-Sicherheit in der Wasserversorgung – Branchenstandard IT-Sicherheit Wasser/Abwasser. *DVGW energie-wasser-praxis* 12 (2016), 134.

[45] *Umsetzungsempfehlungen für das Zukunftsprojekt Industrie 4.0*, Abschlussbericht des Arbeitskreises Industrie 4.0, Vorabversion, 2012.

[46] *Umsetzungsempfehlungen für das Zukunftsprojekt Industrie 4.0*, Abschlussbericht des Arbeitskreises Industrie 4.0, 2013.

[47] *Umsetzungsstrategie Industrie 4.0*, Ergebnisbericht der Plattform Industrie 4.0, 2015.

[48] *UP KRITIS, Öffentlich-Private Partnerschaft zum Schutz Kritischer Infrastrukturen*, Bundesamt für die Sicherheit in der Informationstechnik (BSI), Bonn.

[49] J. Wan, H. Cai, and K. Zhou, *Industrie 4.0: "Enabling Technologies"*, International Conference on Intelligent Computing and Internet of Things (ICIT), 2015, 135–140.

[50] T. Wilschut, T., I. J. B. F. Adan, and J. Stokkermans, *Big Data in daily manufacturing operations*, IEEE Winter Simulation Conference, 2014, 2364.

[51] J. Wübbeke, M. Meissner, M. Zenglein, J. Ives, and B. Conrad, Made in China 2025, The making of a high-tech superpower and consequences for industrial countries, Merics Papers on China, 2016.

Chapter 2

EWave energy management system

Constantin Blanck, Stefan Fischer, Michael Plath, Moritz Allmaras, Andreas Pirsing, Tim Schenk, and Annelie Sohr

Abstract. The EWave energy management system is a decision support system which gives support for an energy and cost efficient operation of water infrastructures. The primary objectives of water supply, which are security of supply and water quality, are extended to include also energy efficiency and automation aspects. The Dorsten-Holsterhausen pilot network provides optimal conditions for testing and evaluating the EWave system in the Rheinisch-Westfälische Wasserwerksgesellschaft mbH (RWW) distribution network. The conditions from the real pilot system to be met during operation and the potentials for improvement to be exploited are translated into various requirements for the development of the EWave system concerning architecture, data interfaces, computational modules and system models.

2.1 Objectives

The operation of water supply systems, such as urban supply networks or long-distance water pipelines, is an extremely complex task that can no longer be fulfilled without the use of modern aids. Nowadays, such supply systems are controlled via central control centers and semi-automated functional processes in accordance with the state of the art. In this context, the operating personnel must ensure that the consumer is guaranteed to be reliably supplied with drinking water of outstanding quality and at a reasonable price.

The aim of the Energy Management System Water Supply (EWave) research project was to develop an innovative energy management system and to pilot it at Rheinisch-Westfälische Wasserwerksgesellschaft mbH as a water supply company with a typical network structure. As a result, energy-optimized operating plans are calculated for the plants operated in the supply system for water production, treatment and distribution. In addition, the fluctuating energy supply from in-house power generation plants and energy procurement from one or more energy supply companies are coordinated, the scope of the optimization calculation was limited by technical and operational restrictions. It was particularly important to ensure quality and security of supply at all times.

The electricity consumption of a water supply company is mainly due to water treatment and water distribution. For this reason, energy optimization focuses on determining the running times or switching times of the network pumps on the one hand and on distributing the required production quantity to the available waterworks on the other. From the point of view of the operating personnel, this results in a highly complex system with interdependencies. Today, this task is almost exclusively performed by experienced operating personnel and requires a very good knowledge of the supply network, in-house power energy generation and the electricity supply contracts concluded with one or more energy supply companies. Together with the changed conditions on the electricity market, shorter contract terms, different tariffs and more varied graduations can be expected in the future. On the one hand, this means that new optimization possibilities arise, but on the other hand, it also means that the operating mode, which up to now has mainly been carried out statically, must become more dynamic and thus the knowledge of the operating personnel accumulated over decades is partially devalued and must be further developed.

The EWave energy management system can serve as a strategic planning aid for the plant management in the first stage of expansion. Understanding the basic considerations of plant operation should be possible, which should be used, among other things, for the short-term formulation of operating instructions.

In a later expansion stage, the use as an operational assistance system (decision support system) should be possible. Irrespective of the expansion stage, a semi-automatic mode of operation with manual post-processing of the calculation results is planned. In a first calculation run, an energy-optimized operation is calculated. This result is presented to the user as a proposal and could either be adopted unchanged or could be easily changed. These changes concerned the boundary conditions, which can be specified by the operator according to the specific situation. After the determination of the new boundary conditions, a new calculation is performed.

The knowledge and experience gained during the research project are transferable to drinking water supply systems with a similar structure. Workshops were therefore held with a broad group of water supply company participants at the beginning and throughout the project duration in order to draw on the experiences of the individual water supply companies with regard to energy management systems and energy-efficient plant operation, to bundle the company-specific requirements and to transfer these into a transferable catalogue of requirements in a further step. By involving other water supply companies at an early stage and in parallel, the acceptance of the approach was already increased beforehand. The planned developments were more specifically geared to practical requirements and the overall chances of exploitation of the project results were thus significantly increased.

2.2 Requirements to the water supply

Despite increasing additional demands on the water supply, the primary objective is to provide the consumer with clean drinking water in sufficient quantity and with appropriate minimum pressure at all times and at a moderate price [3]. Although security of supply is always in the foreground, the efficient use of energy is becoming increas-

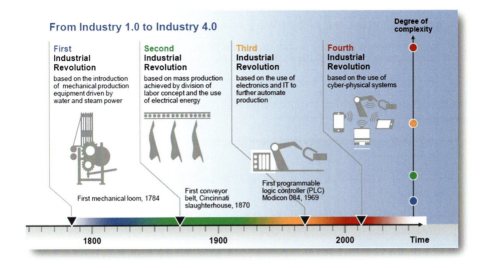

Figure 2.1. Comparison of the four stages of development (Source [1])

ingly important. This steady growth is due on the one hand to the continuously rising energy rates and on the other hand to the energy transition (Energiewende) decided by the German Federal Government. Water supply companies require a relevant amount of electrical energy and thus also release a corresponding amount of carbon dioxide. In order to minimize both the energy required and the climate-relevant CO_2 emissions, in view of climate protection it is therefore also necessary to optimize energy efficiency through continuous development and the use of new technologies.

Due to the increase in digitization and automation, water management is adapting to the growing challenges. As a result of the ever-increasing proportion of digitization, it is possible to view water management systems as a whole in their complexity and its networking. Figure 2.1 shows the four development stages of "Industry 4.0" described by acatech. Analogous the main focus is on merging real and virtual water systems (worlds) into a cyber-physical system (CPS). In this development stage, continuously measured measurement and control variables are permanently available online. In parallel, the available data is evaluated online and further processed by modelling into information that triggers certain measures.

Today, the supply systems are already controlled via semi-automated functional processes and central control centers. At these points of intersection, the operating personnel is responsible for ensuring an inexpensive and reliable drinking water supply. By merging real and virtual water systems it is possible to relieve the operating personnel. On the one hand, the decisions to be made can be supported by the additional information gained. On the other hand, there is the option of permitting system independence through an autonomous cognitive model within a previously defined framework. Such a cognitive model was developed in EWave, which finally allows autonomous action in real time through the perfect fusion of real and virtual world.

The following is a brief summary of the requirements from the perspective of water management and thus of the EWave system to be developed:

- to provide safe drinking water in sufficient quantity and with appropriate minimum pressure at all times and at a moderate price
- efficient use of energy (increase in energy rates and energy transition)

This results in the following requirements from the point of view of digitization and automation:

- consideration of complex and strongly networked water management systems as a whole
- fusion of real and virtual water systems (worlds) to a cyber-physical system (CPS)
- permanent online availability of continuously measured measurement and control variables
- continuous evaluation of data
- option of relieving the operating personnel by the additionally gained information
- autonomously acting cognitive model, system independence within a pre-defined framework

Three energy market issues are also relevant for water supply. Electricity procurement/exchange, atypical grid usage and control energy are intensively considered by many water supply companies. Electricity procurement on the electricity stock exchange is complex and it is not known that it is actively practiced by other water supply companies. As with RWW, the water suppliers procure their electricity with different types of tranche models. This leads to a fixed energy rate for the electricity in the procurement year. Many water supply companies are now participating in atypical grid use. The topic of control energy is more difficult to implement due to the short-term nature of the calls for water supply. There are approaches here, but it is mostly only the emergency power systems of the water suppliers that are offered on the electricity balancing market.

The electricity price for industry and trade is composed, among other things, of the individual unit rate (energy rate) and the individual demand rate. The demand rate is calculated on the basis of the maximum output required in one year. By participating in atypical grid usage, it is possible to reduce the demand rate. The energy rate, however, can be reduced by trading on the electricity stock exchange. RWW currently participates in the atypical grid use with two waterworks, but not with the Dorsten-Holsterhausen waterworks (WHOL). The electricity is not currently traded on the stock exchange, but is procured using a kind of tranche model. This ultimately leads to a fixed energy rate for the electricity in the procurement year.

2.2.1 Electricity stock exchange

The electricity stock exchange is the marketplace where electricity is freely traded. Energy producers can use this trading platform to sell electricity, which in turn is purchased by energy suppliers. Electricity trading is divided into a spot market and a futures market, whereby the spot market is basically divided into two further divisions.

Firstly, there is short-term intraday trading, where electricity is delivered on the same day. In intraday trading, electricity is bought and sold in 15 minute blocks and in hour blocks. Here, it is possible to purchase electricity quotas up to 30 minutes before the planned delivery period. This means that intraday trading is mainly used to reduce shortfalls or surpluses in its own balancing group. The 15-minute electricity trade results in a very dynamic electricity price, which is subject to natural fluctuations.

Secondly, there is day-ahead trading. This includes the purchase and sale of electricity for the entire following day. Here, blocks of hours are traded until 12:00 p.m. for the following day. In contrast to intraday trading, a market clearing price is paid here. This means that the last price at which the required quantity is covered is paid by all market participants. The graph corresponds to a smoothed intraday trading curve.

Due to the flexible energy rate specified by the electricity exchange, this results in a very complex dynamic model caused by the possibility of continuous electricity purchase and sale. The model becomes more complex when it is considered in conjunction with atypical grid usage. The time windows specified by the atypical grid usage, in which the energy should be lowest, reduces the scope of action of the electricity exchange.

2.2.2 Atypical grid utilization

Atypical grid usage is an option in addition to control energy, in order to make a contribution to grid relief and grid stability as final consumer. This is how the final consumers make a significant contribution to the successful energy transition by atypical grid use.

In the case of atypical grid usage, grid operators define quarterly load profiles and peak load time windows based on the electricity consumption of all grid customers in the previous year. High load time windows show the time period with the maximum network load. The aim of atypical grid usage is to ensure grid stability by allowing end users with high loads to call up their maximum loads outside the defined peak load time windows. Due to the shift of the maximum loads of the final consumers, lower loads have to be provided by the network operator. As a monetary benefit, the individual demand rate is reduced from the general maximum load required to the maximum load in the peak load time frame [4].

2.2.3 Control energy

Since only a very small proportion of electrical energy can be temporarily stored in the supply network, the electrical energy generated and required must be balanced. The control energy is used to compensate for unforeseeable power fluctuations. A distinction is made between positive and negative control energy. With positive control energy, the energy required is greater than the energy generated. To compensate for this, additional energy generated has to be fed in or consumers must be switched off. Negative control energy is used when too much energy has been generated. By adding additional energy consumers or reducing generation, the energy required is adapted to the energy generated.

Regardless of positive and negative control energy, the transmission system operator has three basic control energy qualities available to ensure grid stability. The primary, secondary control energy and the minute reserve can be used.

The primary control energy must be fully available within 30 seconds and must be available for at least 15 minutes. In contrast to the secondary energy or the minute reserve, the retrieval is frequency-dependent and independent of the transmission system operator. The supplier of primary energy measures the mains frequency independently. If the mains frequency deviates by 0.01 Hz, the supplier is obliged to compensate for the mains frequency by providing positive or negative primary energy. The primary energy compensates power fluctuations throughout the ENTSO-E (European Network of Transmission System Operators for Electricity) network in Continental Europe.

The power equilibrium of an individual control zone is kept in balance with the aid of secondary control energy. To prevent negative influences from the primary control energy, the secondary control energy is slightly delayed. The full amount must be available within five minutes. In contrast to primary control energy, in the case of secondary control energy, the transmission system operator is responsible for retrieving the secondary control energy. It has direct online access to the technical units operated by the secondary control energy providers to provide secondary control energy. This makes it possible to compensate for power fluctuations as and when required and without delay. The minimum offer size for participation in secondary control energy is 5 MW. Within the same control zone, several plants can be combined to form a pool. It is decisive here that after activation of the secondary control energy at least 1 MW must be available after 30 seconds. The secondary control energy is automatically activated by power-frequency controllers that are used by the transmission system operators [2].

As with primary control energy, secondary control energy is tendered weekly on the website www.regelleistung.net. During the auction, control energy providers specify an output provided and an associated demand rate. The most economical bids are then awarded the contract.

If power fluctuations exceed a time range of 15 minutes, the secondary control energy is replaced by the minute reserve. The minute reserve has a general lead time of 7.5 minutes and must be provided for at least 15 minutes. Unlike secondary control energy, the transmission system operator does not require direct online access to the technical units to provide control energy. The transmission grid operator can, among other things, order the retrieval of the minute reserve by telephone.

In contrast to primary and secondary control energy, the tender for the minute reserve takes place on the previous day and not weekly. As with secondary control energy, the minute reserve, unlike primary energy, is divided into positive and negative control energy [2].

2.3 Pilot network – Dorsten-Holsterhausen

The Dorsten-Holsterhausen waterworks, which already was commissioned in 1927, is one of the main supply areas of RWW in addition to the Ruhr waterworks. Approximately 350,000 people are supplied with drinking water every day through the waterworks.

The Dorsten-Holsterhausen waterworks was selected as the pilot network because it is the easiest to distinguish from all RWW waterworks, including the connected network. The Dorsten-Holsterhausen waterworks supplies the Holsterhausen pressure zone (see Section 2.3.3) and the Gladbeck (PGLA) and Oberhausen-Tackenberg (PTAC) tank system are filled. If, for example, the Mühlheim-Styrum-Ost (WSTO) waterworks were to be considered, three pressure zones would already have to be mapped. Added to this would be the tank systems, which also convey into the zones, and the Mühlheim Styrum-West waterworks (WSTW), which also supplies the same zones. Therefore, the Mühlheim Styrum-West waterworks should also have been modelled. Finally, the model would probably have been at least three times as large.

2.3.1 Water production

The water production of the Dorsten-Holsterhausen waterworks consists of two water catchments. On the one hand there is the Holsterhausen water catchment, which is located in the immediate vicinity of the Dorsten-Holsterhausen waterworks. On the other hand, there is the Üfter Mark water catchment, which is about 10 km north of the waterworks. Both water catchments pump raw water from a deep aquifer, the Halterner sands. Figure 2.2 shows the range of the Halterner sands.

The Halterner sands are composed of fine to medium-grained, partly coarse and silt striped sands, which were deposited about 70 million years ago in the Münsterland chalk basin. In the lower area, the Halterner sands are partially solidified and bounded by quartzite and limestone banks. The upper limit is the Recklinghausen sand marl. In the areas around the Holsterhausen water catchment, the aquifer is overlapped by a groundwater-inhibiting layer dating from the Cretaceous period, the Bottrop marl. Calculations have shown that the groundwater produced by the Holsterhausen water catchment is several thousand years old. With 156 million m^3/a, the Halterner sands provide a sufficiently large groundwater recharge, which to a certain extent benefits the public drinking water supply and to a certain extent the feeding of surface waters [5].

The Holsterhausen water intake consists of 42 vertical wells with the following nominal data:

- nominal flow rate Q: 90 m^3/h
- nominal head H: 90–100 mWS
- nominal electric power P_{el}: 18.5–21 kW

The approved water right for the Holsterhausen water catchment amounts to 25 million m^3/a. The water rights of the Üfter Mark water catchment currently amount to

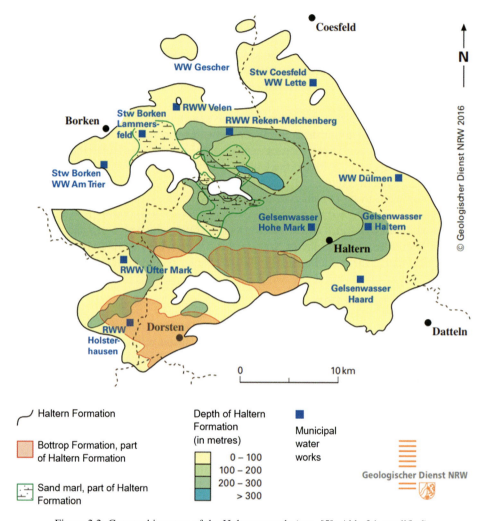

Figure 2.2. Geographic range of the Haltener sands (acc. [5]: Abb. 34, modified)

11 million m³/a. The Üfter Mark well gallery consists of eleven vertical wells with the following nominal data:

- nominal flow rate Q: 100–200 m³/h
- nominal head H: 40–45 mWS
- nominal electric power P_{el} 18.5–30 kW

2.3.2 Water purification

Figure 2.3 shows the treatment scheme of the Dorsten-Holsterhausen waterworks. Due to the geodetic height difference, the raw water of the Üfter Mark has a high

Water works Dorsten-Holsterhausen

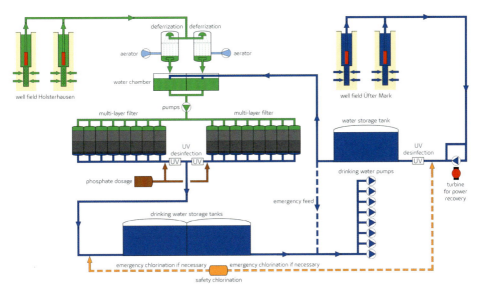

Figure 2.3. Flow sheet water purification, waterworks Dorsten-Holsterhausen

potential energy, which at first is converted into electrical energy at the entrance to the Dorsten-Holsterhausen waterworks via an energy recovery turbine. Before storing the raw water in the Üfter Mark water tank, the water is disinfected with the aid of a UV system.

Due to the geogenic conditions, the raw water from the Holsterhausen water catchment has a higher iron content than the raw water from the Üfter Mark water catchment. The raw water of the Holsterhausen water intake is therefore first aerated via two open oxidizers. In the raw water chambers, the raw water from the Holsterhausen waterworks is then mixed with the raw water from the Üfter Mark water catchment in a specified ratio. The share of raw water in the Üfter Mark capture area can be set to values between 20% and 40% in relation to the total treatment quantity. The oxygen enrichment oxidizes the iron(II) to iron(III) and already partly precipitates in the raw water chambers.

Due to the fact that the raw water of the Holsterhausen capture area has a higher iron content, whereas the raw water of the Üfter Mark capture area has a higher nitrate content, homogeneous raw water is produced by the mixing. The Dorsten-Holsterhausen waterworks operates two parallel filter lines, each with eight multi-layer filters. These are supplied with the raw water from the raw water chambers via six raw water pumps. The individual filters are controlled by a PLC (programmable logic controller) that is linked to the control flap in the filter sequence. This ensures that all filters are supplied with the same quantity in direct current. Depending on the pressure difference of the individual filters, the same amount of water flows through each filter. The filter flushing is based on two interlocking filter flushing criteria. Once a minimum treatment quantity or a maximum differential pressure has been reached, the filters are replenished with a predefined backwash program. After the multi-layer filters, the water is passed through UV reactors and then physically deacidified when it enters the drinking water tank.

2.3.3 Water distribution

The water distribution comprises the drinking water storage tanks, the drinking water pumps that are spatially arranged in the waterworks, and the pipe network. The Dorsten-Holsterhausen waterworks supplies the northern supply area with drinking water with the aid of six fixed speed and two speed-controlled pumps. The supply area includes the following cities and municipalities, represented in Figure 2.4

- Schermbeck,
- Raesfeld,
- Lembeck,
- Wulfen,
- Dorsten,
- Bottrop,
- und Gladbeck.

In addition, the Gladbeck tank system, the Oberhausen-Tackenberg tank system and the Buersche Straße pressure boosting system are located in the supply area. With

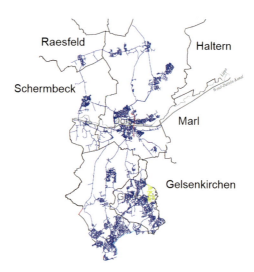

Figure 2.4. Drinking water supply system Holsterhausen [own representation]

a total capacity of approx. 10,000 m^3, which is divided into two tanks, the Gladbeck tank system is RWW's largest tank system. Four fixed speed centrifugal pumps are available at the Gladbeck tank system for conveying from the Gladbeck tank system to the associated supply zone.

To supply the Buersche Straße high zone, the pressure is increased from approx. 4.4 bar to approx. 6.8 bar. This is carried out continuously by one or two pumps in the pressure booster system of the same name.

In addition, the upper Oberhausen zone is partly supplied by the Dorsten-Holsterhausen waterworks via the Oberhausen-Tackenberg tank system. The Oberhausen-Tackenberg tank system has a total capacity of 5,000 m^3, 80% of which is filled at night between 10:00 pm and 6:00 am by Holsterhausen and 20% by the Mühlheim-Styrum-Ost waterworks. Four fixed-speed centrifugal pumps are also available in Tackenberg for conveying from the tank system.

Within the scope of the project, 32 pressure and volumetric flow measuring points were installed in a supply area for the first time. The data recorded serve to gain knowledge of flow rates and pressure conditions in the network and to calibrate the model created in the project.

2.4 Basic concepts of the decision support system

EWave aims to make a significant contribution to digitalization in the water industry by implementing a pilot for the fusion of real and virtual world of water (illustrated above). The focus is on the "digital twin", a virtual model of the real system is continuously updated based on measurements from the real system. On the one hand, the digital twin describes the current state of the system comprehensively and lead to a deep understanding beyond pure measurement ("soft sensors"). On the other hand, it provides a forecast for the following hours and days starting from the current state. The consequences of different operation alternatives become predictable for the operators and they are able to even optimize the operation for the following hours based on the model. The advantage of this approach is that all operation interventions are firstly done virtually without any danger to the facilities and to the security of supply or water quality.

The mathematical optimization of the operation during the forecast period is the core element of EWave. It results in optimal schedules for all controllable pumps and valves. Optimality is meant with respect to energy consumption and energy costs while always giving security of supply the highest priority. Furthermore, operational settings and restrictions, e.g. from maintenance actions, are taken into account.

Figure 2.5 summarizes the EWave optimization approach combining system model and data to calculate optimized schedules.

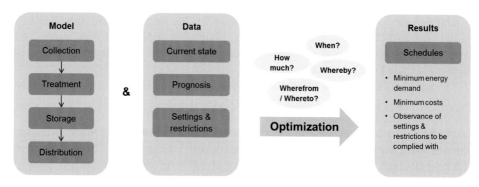

Figure 2.5. Aspects to be addressed by a mathematical optimization

2.4.1 Time aspects

To decide what is the optimal way to operate the real system in the current situation, it is necessary to look into the near future. To make use of the existing scope in water collection, water treatment and water distribution without threatening the security of supply, a prognosis of the water demands for the following hours and any information about which components (pumps, tanks, filters, etc.) will be available is needed. Furthermore, to benefit from dynamic energy pricing and further complex tariff criteria (like atypical network utilization), one additionally must know how energy prices and further restrictions (like power limits) will evolve. Therefore, setting the forecast period is of some importance. On the one hand, the period must be sufficiently long such that it covers the effects of the current operational decisions on the security of supply and energy costs of the following hours. On the other hand, there exists an issue where computational effort increases with the length of the forecast period. Furthermore, forecast quality decreases the farther one looks into the future. The forecast period should, therefore, be sufficiently short to get a fast and reliable support in parallel to operation.

As mentioned, forecast quality decreases over the forecast period. The farther it looks into the future, the less reliable the forecast is. It is therefore not recommended to obey a once optimized schedule for the whole forecast period. For this reason, EWave pursues a cyclic approach with receding time horizon. Optimization is called automatically at periodic intervals. The updated schedules then replace the former ones from this point in time on. The periodic interval must be sufficiently short to be able to react fast to any unforeseen dynamic in the system deviating from the prognosis, but without risking the shortage of time due to too many evaluations.

Finally, the third temporal aspect is the time resolution of the model itself. The smaller the time resolution, the more exact but also more computationally intensive is the model. Operational support is especially affected by the competing objectives optimality and computation time, which should be carefully balanced. One should also consider the time resolution available for the required input date like the prognosis of the energy prices.

The time aspects in EWave are summarized in Figure 2.6.

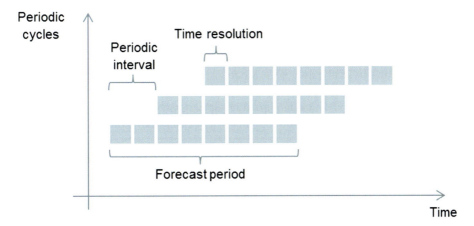

Figure 2.6. Time aspects

2.4.2 System models

The considerations described above for the time aspects are similar for model aspects. Optimality and evaluation time are competing objectives, their results are obviously also influenced by the selection of the model.

Generally speaking, EWave pursues a component-based modeling approach, that means the system model consists of parameterized instances of individual components. These components are collected in a library and describe typical assets in water supply systems like for example pumps, valves and tanks. In these components, both the physical behavior and some simple automation, like for example a local flow or pressure control of a pump, are modeled. The interaction of different components is introduced to the system model by connecting the components.

Compared to classic simulation and optimization applications used in design and engineering of plants and infrastructures, the models used for decision support systems in parallel to the operation must be much more abstract. Long computational time is not acceptable. Given a complex situation, model evaluation has to be fast to allow for a quick and appropriate reaction.

For the model to be applicable in operator support, the following abstraction steps are made in EWave:

- The physical component models are kept simple (e.g. 1D models for pipes) but cover all relevant aspects. This implies setting up an EWave specific hydraulic model library suitable to fulfill the specific requirements.

- The overall system is reduced to a manageable number of components. In particular, concerning the distribution network, only the main lines are considered. The extensive network is accumulated into simple substitute components. In view of a large number of pipes, this step cannot be done manually anymore. A tool to support network abstraction is necessary.

- Optimization repeatedly evaluates the model for different operational settings.

To do this in an efficient manner, further mathematical approaches to model reduction are applied.

As in all model-based applications but in particular relevant for abstract, substitute models with non-physical parameters, model calibration is of huge importance. Calibration is mainly done based on recorded, historic data from real system operation. If the models are continuously used in operator support parallel to real system operation, one should also think about an automated, cyclic re-calibration during operation.

2.4.3 Coupling simulation and optimization

As described above, the evaluation of the system model as part of an optimization of the system operation is the core element of EWave. Before discussing the implications, it is important to clarify some terms in the context of simulation and optimization of operations:

- A dynamic simulation is an evaluation over time of a model that describes the time-varying behavior of a system.
- Different operation alternatives can be evaluated based on a dynamic simulation of the model with different boundary conditions (i.e. time-varying loads, settings, restrictions) and followed by a comparison of the simulation results.
- An optimization algorithm generates, evaluates, compares and prioritizes operation alternatives in an automated manner with the aim to find the optimal solution.

Integrating an optimization into EWave implies the following requirements:

The optimization requires an initial state which reflects the current situation in the networks and which is feasible according to the model equations. In particular, the pressure distribution in the network has to be reproduced by the initial state. Measurements from the real system like filling levels and flows do not suffice to define this initial state. Pressure measurements are available to an only very limited extent, if at all. In EWave, this problem is solved by running a simulation before calling the optimization. The simulation will run over some lead time and adapt to measurements from the real system and thus will calculate the current initial state for the optimization.

The results of the optimization are schedules for all controllable pumps and valves. As said before, the optimization model must be abstract to keep computation time within acceptable limits. Therefore, it may be of use to run a more detailed system simulation for the optimized schedules after the optimization run. By doing this, optimized schedules can be verified with respect to their feasibility, implications to the system behavior can be predicted in more detail and further key performance indicators can be calculated. EWave has, therefore, implemented a detailed simulation to follow the abstract optimization.

The main sequence in EWave is an optimization preceded and followed by a simulation as shown in Figure 2.7.

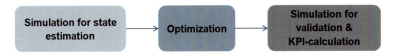

Figure 2.7. Main sequence in a single cyclic execution

2.4.4 Prognosis

Evaluating the model over the forecast period requires having a prognosis for all time-varying input data for this period. In detail, these are

- **Prognosis of demands.** A prognosis of time-related and locally resolved water demands over the forecast period is needed. The demand prognosis is considered as an integral part of the EWave system. But integrating an external demand prognosis is also possible.
- **Prognosis of energy prizing.** Tariff characteristics like unit rate, demand charge, power limits, the time frame for atypical network utilization, etc. must be known. If applicable, a prognosis of dynamic energy prices is needed. Interfaces for reading an appropriate prognosis and models to consider it in optimization are implemented in EWave. The prognosis itself is not a part of EWave.
- **Prognosis of own-energy generation.** Own-energy generation allows a water supply company to reduce costs for purchasing energy from other energy suppliers. Interfaces for reading an appropriate prognosis and models to consider it in optimization are implemented in EWave. The prognosis itself is not a part of EWave.

2.4.5 Boundary conditions

When trying to find the optimal operation schedules, of course, all given restrictions have to be respected. EWave takes the restrictions into account. Parametric specifications are part of the component models. In detail, these are:

- technical restrictions, like operating ranges of pumps, treatment plants and tanks, restrictions for switching pumps, etc.
- legal requirements (short and medium term), e.g. considering water extraction from wells
- operation plans, service & maintenance plans, e.g. maintenance for pumps, cleaning of tanks, flushing filters
- existing local automation, not to be replaced by EWave optimized schedules, e.g. pressure control by booster pumps
- further restrictions, e.g. mixing ratios to keep up water quality

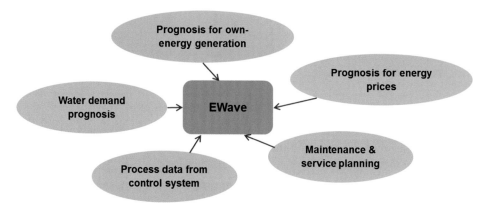

Figure 2.8. Data environment

2.4.6 Data access

Both for setting up the EWave system and for running it during operation, the access to various data from different sources is essential.

In the first place, for setting up the model's various static data (characteristics of technical components, geometries of tanks and pipes, altitudes) are necessary.

For calibration of both the simulation and optimization models and the model for the demand prognosis access to historic operational data over long periods of time (months to years) is desirable. The data originate from the control system or from external sources (e.g. water demand, weather data).

For running the EWave system in parallel to the operation, access to current measurements and operating notifications from the control system is necessary. Also, access to the preceding hours is needed to be used for the simulation-based calculation of the current state.

For external demand prognosis, the prognosis of energy prices and the prognosis of own-energy generation, if applicable, interfaces have to be implemented, since this data usually is not available in the control systems.

For operation planning influencing the availability of components of the model, an interface to an existing maintenance & service planning system is conceivable, but is not part of a first version of the EWave system.

A summary of the required access to data, when used in operation, is given in Figure 2.8.

Bibliography

[1] Siemens AG: Siemens Blog – Better Big Data analytics for product performance intelligence. Online-Blog. https://community.plm.automation.siemens.com/t5/Digital-Transformations/Better-Big-Data-analytics-for-product-performance-intelligence/ba-p/338044. 2016

[2] Bundesnetzagentur (2011): "Beschluss – BK6-10-098"

[3] DIN 2000 (2017): Zentrale Trinkwasserversorgung – Leitsätze für Anforderungen an Trinkwasser, Planung, Bau, Betrieb und Instandhaltung der Versorgungsanlagen, Beuth-Vertrieb GmbH, Berlin.

[4] DIHK (2015): "Faktenpapier atypische Netznutzung", DIHK – Deutscher Industrie- und Handelskammertag Berlin | Brüssel und VEA – Bundesverband der Energie-Abnehmer e.V. Hannover

[5] Wrede, V. (2016): Schiefergas und Flözgas. Potenziale und Risiken der Erkundung unkonventioneller Erdgasvorkommen in Nordrhein-Westfalen. – Scriptum 23: 129 S., 42 Abb., 8 Tab.; Krefeld.

Part II

Theoretical aspects

The new decision support system follows a model-based approach. A mathematical model of the entire system, comprising water recovery, purification and distribution is used in various abstraction grades for simulating the physical behavior in the specific operating mode and for optimizing this plant operation. The modeling of the water supply system is component-oriented, i.e. the network is constructed from a catalog of components which are instantiated, assigned parameters and linked to one another. For water distribution, the library comprises, for example, the components of pipe, pump, valve and tank. Depending on the application, the physical behavior is simulated on a different abstraction level. A simulation run, for example, uses a more detailed level than an optimization run. The common feature, however, is the general procedure when constructing the mathematical model. Each component contributes a specific set of equations to the overall system. These can not only be algebraic equations, but also differential equations. Coupling conditions among the components, which simulate principles such as the mass conservation and equality of pressure, complete the set of equations. For a detailed perspective of the simulation, a hydrodynamic model is created. The essential physical principles that are considered here are mass and momentum conservation. The type of the resulting equation systems is greatly dependent on the level of abstraction. In the event of a hydrodynamic simulation, differential-algebraic equation systems (DAE systems) are received that can be solved with DAE solvers, e.g. on the basis the implicit Runge-Kutta or Rosenbrock-Wanner procedure. In the simplest case, the considerably more abstract models that are used for an optimization lead to linear problems, often combined with an integer component, so that procedures from the mixed-integer linear programming (MILP or MIP) family are used.

Chapter 3
Demand forecast

Patrick Hausmann

Abstract. The following chapter deals with the statistical demand forecast for the waterworks in Dorsten-Holsterhausen, which was developed during the EWave research project. Basic principle for the forecast development was the assumption that demand levels on similar days (e.g. every Monday in February) are equal. Therefore the historical demand levels were analysed and classified depending on time, day and year (clustering). The historical data was provided by the water supply company RWW as hourly measured time series values from a period of several years. After the analysis, demandchanges were fitted with first order polynomials (regression lines). Thereby forecast values with a time resolution of one hour could be calculated. To avoid discrete transitions between forecast values and increase their time resolution a spline interpolation was carried out afterwards. It allowed artificial time resolutions of up to one second and guaranteed continously differentiable forecast values. In the end forecast results were compared with current measures at RWW to determine and analyse accuracy and quality of the forecast program.

3.1 Introduction

3.1.1 Necessity of a demand forecast

Aim of the joint research project EWave is the operational optimization of water supply systems regarding energy efficiency and security of supply. For this purpose simulation and optimization tools are developed for which water demand is an important input value.

Since an effective optimization also needs to consider future values a water demand forecast tool is required. Knowing about future demand values allows the system to prepare for specific situations and calculate optimal control parameters in advance. A typical example is the usage of reservoirs: If the future water demand is known, high-lying reservoirs can be filled to reduce the usage of pumps during high demand periods. Besides providing demand values for energetic optimization, these values can also be utilized when considering economical aspects, such as electricity rates. The initial situation in the beginning of EWave was a qualitative forecast of future demand levels based on long lasting experiences of the waterworks personnell.

This approach has at least two major disadvantages: Firstly, it is not well founded and based entirely on subjective experiences. Therefore it is also very dependent on the individual staff and not future proof. Secondly, it is in no way quantifiable and thus not usable for the simulation or optimization tools which are employed for EWave.

By developing a numerical, statistical based demand forecast, well grounded values can be calculated to be used for the named tools and also be filed for potential future analysis.

The following sections are about model approach, data clustering and parametrization as well as program flow and accuracy of the developed demand forecast.

3.1.2 Concept, model approach, and classification

In the context of this chapter the term "demand" or "water demand" means the total water output of the waterworks delivered by the drinking water pumps and typically measured in m^3/h. The demand values provided by the water supply company are hourly measured and represent the average water usage of the affiliated supply zone over one hour.

Concept. The demand forecast is based on the assumption that demand levels are comparable on equal days. More accurate it is assumed, that the change of demand from one point in time to another is related to the current date and time. For example: The demand on a Tuesday in March at 14:00 is $2,900\,m^3/h$. At 15:00 it changes to $3,000\,m^3/h$ which equals an increase of 3.4%. It is now assumed, that this rate of increase for this time of day applies to every Tuesday in March. So if, on another Tuesday in March at 14:00, the demand is e.g. $3,200\,m^3/h$ the future demand for 15:00 will be calculated as $3,200\,m^3/h + 3.4\% = 3,310\,m^3/h$.

Model approach. To determine the rates of increase for each hour, day and month, historical data provided by the water supply company has to be evaluated. Therefore extensive time series sheets of water demand levels from several years are provided by RWW.

Firstly, these time series have to be clustered. The clustering is based on the time of day, day of the week and month. So eventually there are several historically recorded demand levels for every possible time of the year stored in clusters. Every such cluster is called a time class (e.g., 14:00–15:00, Tuesday, March). Sundays and holidays are regarded as the same time class.

Afterwards, every possible change from one time class to another has to be identified, before the rates of increase (or decrease) for these time changes are determined.

A forecast value can now be calculated by providing an initial demand value. This inital value is usually provided by the EWave system. The second forecast value is then calculated by restarting with the first forecast value and so on. Since the resolution of the underlying time series data is only one hour, the resolution of the forecast results is also one hour.

After every demand level for the desired range has been calculated the time resolution can be artificially increased by using a spline function. With this extension, the time resolution of the forecast can be adjustable at will, with a maximum resolution of one second (see Figure 3.4).

Classification. Demand levels of the forecast are calculated utilizing first order polynomials (regression lines). These polynomials are parameterized using the historical time series. As such, the model can be categorized as a linear least square regressionmodel. Since the selection of polynomial coefficients is time dependent it is also a local model with time dependent function parameters [1].

3.2 Demand forecast

The following sections describe the demand forecast in detail. At first the process of clustering the time series data is specified. Following, the model parametrization based on this data is explained. After the program structure is clear, the program flow is clarified and results and accuracy are evaluated.

3.2.1 Data clustering

In preparation of the forecast model the provided time series data has to be clustered (see Figure 3.1). This is necessary in order to parametrize the regression lines for demand changes later on. In general there are three main classes for every point in time:

- M01 – M12 Month class.
  January, February, ..., December
- D01 – D07 Day class.
  Holidays are threated as Sundays. Monday, Tuesday, ..., Sunday/Holiday
- T01 – T24 Time class.
  00:00–01:00, 01:00–02:00, ..., 23:00–24:00

After clustering the data, the possible changes between different classes (e.g. from 23:00-24:00 on a Monday in March to 00:00-01:00 on Tuesday in April) have to be identified. Since the changes in times of day are always the same, only changes in the D- or M-class have to be considered. There are in sum 18 possible changes for days (CT1) and 24 for months (CT2) (staying in the same class, e.g. from January to January is also defined as a "change", see also Tables 3.1 and 3.2).

Changes between workdays are threated as regular, changes from workdays to holidays or Sundays as irregular. It has to be noted that there is significantly more data availabe for regular changes than for irregular, simply because of their frequency. This has to be considered when parameterizing the model.

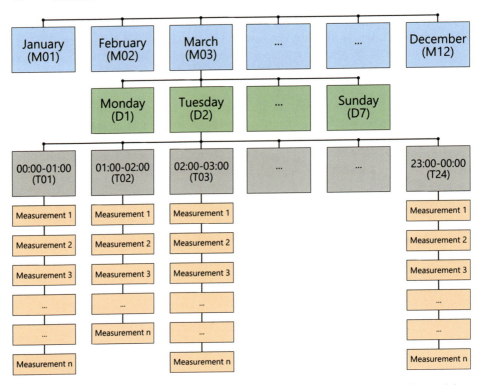

Figure 3.1. Clustering of hourly averaged demand-values according to the respective date and time of measurement (example)

3.2.2 Model parametrization

After the data clustering is completed, the identification of model parameters can begin. Therefore, the demand levels at two neighbouring points in time are evaluated. Since the time series data has already been clustered, there should be a variety of available data values for both of the points. The rate of change is now approximated using a first order polynomial (see Eq. 3.1).

$$q_i(q_{i-1}) = a_1 q_{i-1} + a_0 \tag{3.1}$$

- q_i – Demand value at time t_i, e.g., 14:00–15:00
- q_{i-1} – Demand value at time t_{i-1}, e.g., 13:00–14:00

As already mentioned in the section about data clustering, the available data for irregular time changes (e.g. change from Wednesday to a holiday) is very limited. For some events there is only one data point available. With only one pair of values to approximate, the a_0 parameter of the calculated polynomial is set to zero, which is inconsistent with the other regression lines (see Figures 3.2 and 3.3). In it's current state the model does not consider this and calculates the demand values with incon-

Change	CT1	Change	CT1	Change	CT1
MO-TU	1	SU-MO	6	MO-SU	13
TU-WE	2	SU-TU	7	TU-SU	14
WE-TH	3	SU-WE	8	WE-SU	15
TH-FR	4	SU-TH	9	TH-SU	16
FR-SA	5	SU-FR	10	FR-SU	17
		SU-SA	11	SA-SU	18
		SU-SU	12		

Table 3.1. Possible changeovers of days

Change	CT2	Change	CT2	Change	CT2	Change	CT2
JAN–JAN	1	JUL–JUL	7	JAN–FEB	13	JUL–AUG	19
FEB–FEB	2	AUG–AUG	8	FEB–MAR	14	AUG–SEP	20
MAR–MAR	3	SEP–SEP	9	MAR–APR	15	SEP–OCT	21
APR–APR	4	OCT–OCT	10	APR–MAY	16	OCT–NOV	22
MAY–MAY	5	NOV–NOV	11	MAY–JUN	17	NOV–DEC	23
JUN–JUN	6	DEC–DEC	12	JUN–JUL	18	DEC–JAN	24

Table 3.2. Possible changeovers of months

sistent coefficients in this case. But since the occurance of these cases is very rarely this problem is currently neglected.

After all parameters for every possible change have been identified they are stored in a lookup table for later use. With 24 time changes of 18 possible day changes (see Table 3.1) and 24 possible month changes (see Table 3.2) the model ends up with 20,736 coefficients for the regression lines.

3.2.3 Artificial increase of time resolution

The demand forecast primarily calculates discrete, hourly values. This can lead to some difficulties when using them in following programs, e.g. network simulation, because the transition between two values is not continously differentiable.

To fix this issue, a spline interpolation of the primarily calculated values is applied which also artificially increases the time resolution of the forecast. Figure 3.4 exemplarily shows the primary results of the forecast (blue) plotted against the corresponding spline interpolation (orange).

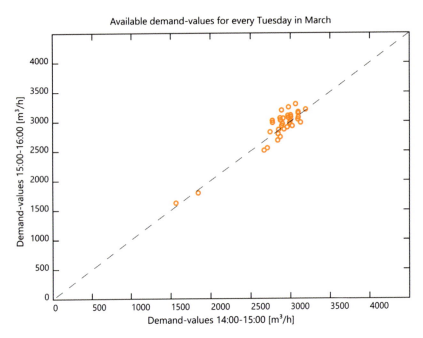

Figure 3.2. Regression line for an hourly demand-change with sufficient data points (regular change)

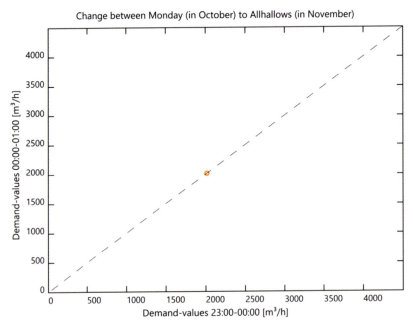

Figure 3.3. Regression line for an hourly demand-change with insufficient data points (irregular change)

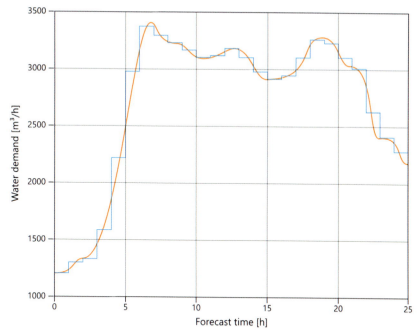

Figure 3.4. Hourly forecast results and spline interpolation

Input values for the spline interpolation are the hourly forecast values $\bar{d}_i, i = 1, \ldots, h_{max}$ for the desired forecast range with a maximum of h_{max} values (e.g., 24). Firstly, these average flow rates over the time period $t \in [t_{i-1}, t_i]$ are integrated to get total demands per hour:

$$\bar{d}_i = \frac{1}{t_i - t_{i-1}} \int_{t_{i-1}}^{t_i} d_i(t) dt \tag{3.2}$$

In order to identify $d_i(t)$ a cubic spline approach is applied:

$$d_i(t) = a_i + b_i(t - t_{i-1}) + c_i(t - t_{i-1})^2 + d_i(t - t_{i-1})^3 \tag{3.3}$$

Coefficients a_i, b_i, c_i, d_i are computed in order to generate a twice continous differential function:

$$d(t) = d_i(t), \quad \text{if} \quad t \in [t_{i-1}, t_i], \tag{3.4}$$

which conserves the total demand values given in Eq. 3.2. The three remaining boundary conditions are chosen as $d'(0) = d''(0) = d'(h_{max}) = 0$.

3.2.4 Program flow

The demand forecast program is written in MATLAB and later compiled as an executive file to be launched by the EWave system.

To start a forecast, the program has to be called with values for the current date and time, an initial demand value, the desired duration of the forecast and time resolution. These values are usually provided by the EWave system. After the inputs have been validated, the program starts the processing of the transferred date and time values (see Figure 3.5).

Date and range evaluation. First step of the date evaluation is a check whether the transferred date is in a leapyear. If this is true, February is defined with a length of 29 days. After that, the program also checks whether the current day is the last day of the year. The next step is evaluating the forecast range and defining a date list to know which days in the current and following year are holidays. By doing this the program also defines which class changes will occur.

Selection of coefficients and demand forecast. The selection of polynomial coefficients is embedded into a for-loop whose length is the forecast range (in hours). For the choice of coefficients the program consults the named date list to define which month, day and time classes occur during the forecast range. Since the coefficients for every class change have already been calculated and stored in a lookup table, the program now can deduce a list of the currently required coefficients. Next step is the demand forecast. Based on the transferred initial demand level the program calculates future hourly demand levels by using the polynomial coefficients which were just mentioned. The calculation is recursive, which means that the calculation of a new value is always based on the value prior to it (see Eq. 3.1). Because of that, the desired demand range should not be too large since it gradually decreases the forecast accuracy. Finally the forecast result is stored in a list to be used for arificial increase of time resolution.

Artificial increase of time resolution. Since the forecasted demand levels only have a time resolution of one hour, the programs offers a feature to articially increase it's time resolution. This spline can now be evaluated at random points in time while still being consistent with the original forecast values.

3.2.5 Results and accuracy

Quality und accuracy of the demand forecast. The overall performance of the demand forecast is very satisfying. Looking at the results (see Figure 3.6), an accurate qualitative course of the forecasted demand levels is noticeable. The quantitative accuracy is also satisfying though there is still room for improvement.

To evaluate the program, forecasted values are compared to measured demand values provided by the water supply company. Evaluation parameters are the relations in % and absolute difference in m^3/h between forecast and measured values. The results are as follows:

- Accuracy of flow rate prediction: $\approx 93.0\%$
- Accuracy of overall fluid quantity: $\approx 182.3\, m^3/h$

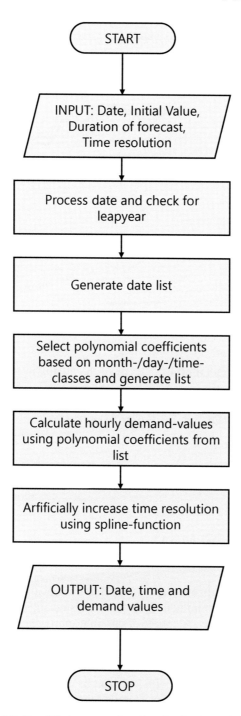

Figure 3.5. Simplified program flowchart for the demand forecast

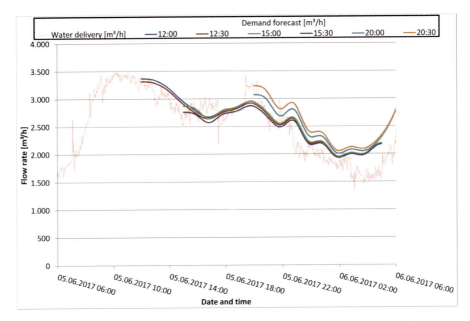

Figure 3.6. Forecast results plotted against measured water demands at RWW

Explanation of error calculation. Basis of the quality evalutation is an error calcuation. For this, the root mean square (RMS) deviation from forecast values to measured values is calculated both for flow rate and overall fluid quantity.

The used formulas (Eq. 3.5, Eq. 3.6) to calculate this error are mentioned below.

$$\bar{x}_{RMS} = \sqrt{\frac{1}{24} \sum_{i=1}^{24} \left(\frac{q_{prog,i}}{q_{mes,i}} - 1 \right)^2} \qquad (3.5)$$

$$\bar{x}_{RMS} = \sqrt{\frac{1}{24} \sum_{i=1}^{24} \left(q_{prog,i} - q_{mes,i} \right)^2} \qquad (3.6)$$

Basic approaches to improvements. Inaccuracies of the demand forecast occur due to diverse reasons. An influential factor are systematic errors e.g. due to inaccurate polynomial coefficients. A couple of possible approaches to improve the forecast accuracy are listed in the following:

- Elimination of inconsistent polynomial coefficients.

  Inconsistent polynomial coefficients can be eliminated by e.g. replacing them with coefficients derived from a similar class change.

α is the adaptation coefficient and factors f^+, f^- are choosen as

$$f^+ = \frac{1}{2}(\text{sgn}(d(t)) + 1) \; ; \quad f^- = \frac{1}{2}(1 - \text{sgn}(d(t))) \, , \quad (4.7)$$

where $\text{sgn}(\cdot)$ is the signum function. A backflow preventer can be modeled by a control valve using $d(t) = Q(t)$. Then the valve gets closed when $Q(t)$ becomes negative.

Pump. A pump is described by a strictly monotone decreasing relationship between delivered flow $Q(t)$ and pressure difference $d_H(t) = H_R(t) - H_L(t)$. It is assumed that this relationship can de modeled by $d_H(t) = \alpha_0 - \alpha_r Q(t)^r$ with positive constants $\alpha_0, \alpha_r, r > 1$. For implementation, the equivalent characteristic curve is preferred:

$$Q(t) = i_o(t) \left(\frac{\alpha_0 - d_H(t)}{\alpha_r} \right)^{\frac{1}{r}} . \quad (4.8)$$

The pump can be switched on or off by $i_o(t) \in \{0, 1\}$. For $i_o(t) = 1$ the curve should be extrapolated via $Q = 0$ for $d_H > \alpha_0$ and $Q = Q_{max} = (\frac{\alpha_0}{\alpha_r})^{1/r}$ for $d_H < 0$.

If the revolution speed $n(t) \in [n_{min}, n_{max}]$ of the pump can be regulated, it is assumed that it is possible to interpolate between two chracterstic curves with coefficients $\alpha_0^1, \alpha_r^1, r^1$ and $\alpha_0^2, \alpha_r^2, r^2$ valid for minumum and maximum speed respectively. For coefficients α_0 and r linear interpolation is applied. Coefficient α_r is then computed by $\alpha_r = \frac{\alpha_0}{Q_{max}^r}$ where Q_{max} is again linearly interpolated from the maximum possible flows Q_{max}^1, Q_{max}^2 for speeds n_{min} and n_{max}, respectively.

Control pump. The speed of a control pump is regulated similar to the opening degree of a control valve. Depending on the difference $d(t) = H^t - H(t)$ of a pressure target H^t and corresponding state $H(t)$ at the pump outlet or another location within the network behind the pump, the speed $n(t)$ is adapted according to

$$n'(t) = \frac{d(t)}{\alpha} \left(f^+(n_{max} - n(t)) + f^-(n(t) - n_{min}) \right) ; \quad n(t_0) = n_0 . \quad (4.9)$$

Factors f^+, f^- are chosen according to (4.7).

Pipe. In contrast to connections, valves and pumps, in pipes space dependent flow $Q(x, t)$ and pressure head $H(x, t)$ are considered. One-dimensional flow equations describing conservation of mass and momentum of water along a pipe element are denoted as water hammer equations and read, [1]:

$$\frac{\partial H}{\partial t} + \frac{a^2}{gA} \frac{\partial Q}{\partial x} = 0 , \quad (4.10)$$

$$\frac{\partial Q}{\partial t} + gA \frac{\partial H}{\partial x} = -\lambda(Q) \frac{Q|Q|}{2DA} , \quad (4.11)$$

Beside the coupling conditions (4.1), (4.2), for each edge i some state equations must be defined. In most cases $Q_L^i(t) = Q_R^i(t) =: Q^i(t)$ and a nonlinear relationship $f(Q^i(t), H_L^i(t), H_R^i(t)) = 0$ hold. E.g., function f can be given by a characteristic curve of a pump.

Flow or pressure controlled valves and pumps are supplemented by an ordinary differential equation (ODE) and pipes are modeled by partial differential equations (PDEs). In summary, one gets a large system of coupled PDEs, ODEs and linear as well as nonlinear algebraic equations, denoted as PDAE problem. Corresponding examples are well known from electric circuit simulation, see [4].

In Table 4.1 all types of considered network elements are given. The corresponding equations are introduced in the next section. Moreover, to each element a specific energy demand, depending on the hydraulic state can be assigned. This is important for optimizing the overall energy demand of the whole network.

4.2.2 Modeling equations

Connection. The most simple type of an edge is a connection. Its equations are given by $H_L(t) = H_R(t)$ and $Q(t) = Q_L(t) = Q_R(t)$. The upper index denoting the number of the corresponding edge is omitted in this section.

In some cases it may be necessary to require a pressure loss H for a connection:

$$H_L(t) = H_R(t) + H(Q). \tag{4.4}$$

This pressure loss may depend on the flow Q by a given characteristic curve, such as: $H(Q) = \alpha_0 + \alpha_1 Q + \alpha_r Q^r, \alpha_0, \alpha_1, \alpha_r, r \in \mathbb{R}$.

Valve. At a valve inflow equals outflow, too: $Q_L(t) = Q_R(t)$. But it leads to another type of pressure loss:

$$s^2(t)(H_L(t) - H_R(t)) = \tilde{\zeta} |Q| Q. \tag{4.5}$$

Here, parameter $\tilde{\zeta}$ is given by $\tilde{\zeta} = \frac{\zeta}{2gA^2}$ with pressure loss coefficient ζ and gravitational constant g. The opening degree $s(t)$ is the ratio of opened to total cross section of the valve, i.e. $s(t) = \frac{A_s(t)}{A}$. It may vary with time.

Control valve and backflow preventer. In case of a control valve, $s(t)$ is adapted depending on a given flow target Q^t or targets for input pressure H_L^t or output pressure H_R^t. It is assumed that the rate of change $s'(t)$ is proportional to the difference $d(t)$ given between target value and the current state, i.e. $d(t) = Q^t - Q(t)$ or $d(t) = H_R^t - H_R(t)$ or $d(t) = H_L(t) - H_L^t(t)$. Moreover, $s(t)$ must be bounded, i.e. $s(t) \in [s_{min}, s_{max}]$, where $s_{min} = 0$, $s_{max} = 1$ is choosen usually. These requirements can be fullfilled by following ordinary differential equation and corresponding initial condition at time t_0:

$$s'(t) = \frac{d(t)}{\alpha} \left(f^+ (s_{max} - s(t)) + f^- (s(t) - s_{min}) \right); \quad s(t_0) = s_0. \tag{4.6}$$

These elements should allow a mapping of the real water work and water supply network to the mathematical model in order to represent main processes related to hydraulics and energy sufficiently accurate.

In the Section 4.2 the mathematical model for each network element and the coupling conditions are given. Based on this component library, Sections 4.3, 4.4 deal with the numerical solution of resulting equations and the implementation of appropriate simulators. Simulator TWaveSim is used for detailed simulation of hydrodynamic behavior of the whole network while the other simulator Anaconda is mainly applied for continuous optimization. Finally, an example demonstrating the principle procedure is introduced in Section 4.5. Application to a real water work and pressure zone is discussed in in the next chapter of this book.

4.2 Water supply network modeling

4.2.1 Component-based network approach

The entire hydraulic model is based on state variables pressure head H and volume flow Q of water. It is assumed that each network element, like a pipe, pump etc. can be considered as an edge and the edges are connected via nodes. Thus, the whole model is represented by a directed graph [5, 9].

The state variables of an edge i depend on space x and time t: $H^i(x,t)$, $Q^i(x,t)$. Space dependency is resolved only within pipes. For all other elements just inlet (L=left) and outlet (R=right) are distinguished: $H_L^i(t)$, $H_R^i(t)$, $Q_L^i(t)$, $Q_R^i(t)$. At nodes it is assumed that pressure heads of all connected edges are identical. When, for example elements i, j are connected via their outlets to the inlet of element k, the coupling conditions at that node read:

$$H_R^i(t) = H_L^k(t), \quad H_R^j(t) = H_L^k(t). \tag{4.1}$$

Moreover, at each node mass balance is regarded: The sum of flows must be zero in any node. For the aforementioned example this gives:

$$Q_R^i(t) + Q_R^j(t) - Q_L^k(t) + Q_s(t) = 0. \tag{4.2}$$

Here, $Q_s(t)$ is an additional source or sink term for adding or extracting water at that node. Sometimes, at a node a specific pressure $H_s(t)$ is prescribed. Then, (4.2) must be replaced by

$$H_R^i(t) = H_s(t). \tag{4.3}$$

A tank is treated as a node, too. In that case the node has storage capacities and equation (4.2) is replaced by a differential equation, which is given in the next section.

Chapter 4
Hydraulic modeling and energy view

Gerd Steinebach and Oliver Kolb

Abstract. In this chapter the hydraulic modeling of a waterworks and the drinking water distribution network is considered. At first, network elements are introduced, which allow the mapping of the real water supply network to the mathematical model. For each network element a mathematical description is given and coupling conditions are defined. Moreover, each element can be provided with an elctrical power requirement, in order to compute the overall power demand for the whole network.
For the resulting mathematical model two simulators are introduced. TWaveSim can only be used for simulation, while Anaconda is also suitable for continuous optimization. On the other hand, TWaveSim is more accurate for the simulation purpose and must be applied first, to compute initial values that are required for Anaconda.
Finally, a test example is constructed and simulated. This example includes most of the introduced network elements and is used to show all mathematical equations in detail and to compare the two simulators.

4.1 Introduction

Physical description of main processes within water works and the water supply networks is an important issue of the whole EWave project. Based on this description, an appropriate simulator can be set up in order to compute hydraulic and energetic states at different times. Optimization modules need the current states as initial conditions for computing optimized operation of pumps and valves. Using these pump and valve schedules and forecasted water demands, the simulation module again computes hydraulic and energetic states of the whole system for the next 24 hours.

The simulator is based on a component-based hydraulic modeling approach, aspects of water quality are not considered. In a first step, basic network elements are identified, see Table 4.1.

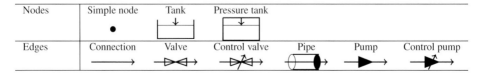

Table 4.1. Network elements

- Autoregression.

  The current forecast model calculates future demand levels based on the average demand of similar days in the past. An improvement of the model could be achieved by also considering newly measured demand levels and thus enhancing the dynamic of the model.
- Higher resolution of time series data

  The model was parametrized using historical time series data with a time resolution of one hour. A higher resolution of the fundamental data set allows for a more accurate parametrization and thus promises a higher forecast accuracy.

3.3 Summary

In this chapter the statistical demand forecast for the waterworks in Dorsten-Holsterhausen was described in detail. It was developed during the EWave research project, partially for RWW which is also the operating company. During the project the need for a demand forecast arose because realistic input values for simulation and optimization models were needed. Furthermore, simulations as well as optimizations should already consider future demand situations to calculate optimal operating plans for the waterworks personell beforehand. Before the the model was developed, historical time series data which were measured by RWW over a period of several years were analysed. This data was grouped into different classes (Clustering). For example the demand values between 14:00 and 15:00 of every Monday in July were assigned to one class. Basis of the forecast model is the assumption that courses of demand are equal on similar days. Therefore the clustered time series data was analysed and the changes in demand over e.g. one day were approximated using first order polynomials (regression lines). The time resolution was one hour which matches the time resolution of the historical data.

Bibliography

[1] K. Specht and H. Rinne, *Zeitreihen – Statistische Modellierung, Schaetzung und Prognose*, Verlag Franz Vahlen, 2002.

with $x \in [x_L, x_R]$, $t > t_0$. For simplification, cross sectional area A, diameter D of the pipe and sound velocity a are supposed to be constant.

For turbulent flow, friction coefficient λ can be approximated by the Swamee-Jain-approximation of the Colebrook and White formula, [15]:

$$\lambda(Q) = \frac{0.25}{\left(\log_{10}\left(\frac{k}{3.7D} + \frac{5.74}{Re^{0.9}}\right)\right)^2} \quad (4.12)$$

with roughness parameter k of the pipe. The Reynolds number is given by $Re = \frac{4|Q|}{\pi D \nu}$ with kinematic viscosity ν.

Equations (4.10), (4.11) must be supplemented by two physical boundary conditions which realize the coupling according to (4.1), (4.2). Additionally, two numerical boundary conditions are required, see Section 4.3.2.

Simple node. As noted in Table 4.1, there are different types of nodes. Simple nodes are not connected to own state variables. They are only needed for the coupling of different edges according to the equations (4.1) and (4.2).

Tank. Tanks and pressure tanks are considered as nodes with an own state variable $H_T(t)$. Regarding a tank, coupling condition (4.2) is substituted by the equation for mass conservation. In case of constant base area A of the tank, mass conservation is given by the initial value problem

$$H'_T(t) = \frac{1}{A}(Q_T(t) + Q_s(t)), \quad H_T(t_0) = H_0. \quad (4.13)$$

$Q_s(t)$ is an external inflow or outflow and $Q_T(t)$ is the flow caused by the connected edges. Using again the assumptions of equations (4.1), that the tank is connected to elements i, j via its outlets and to the inlet of element k, it follows $Q_T(t) = Q_R^i(t) + Q_R^j(t) - Q_L^k(t)$ and coupling conditions (4.1) must be complemented by

$$H_R^i(t) = H_T(t). \quad (4.14)$$

Alternatively, it can be assumed that each in- or outflow to the tank causes a pressure loss. In that case, the flow caused by each edge is modeled exactly like the flow through a valve, see equation (4.5), and conditions (4.1, 4.14) must be replaced by:

$$\tilde{\zeta} Q_R^i(t)|Q_R^i(t)| = H_R^i(t) - H_T(t), \quad (4.15)$$
$$\tilde{\zeta} Q_R^j(t)|Q_R^j(t)| = H_R^j(t) - H_T(t), \quad (4.16)$$
$$\tilde{\zeta} Q_L^k(t)|Q_L^k(t)| = -(H_L^k(t) - H_T(t)). \quad (4.17)$$

When the tank is empty no outflow is possible. This can be modeled by changing valve equations (4.15, 4.16, 4.17) to those of control valves, see (4.5, 4.6). The control valves must close, when $H_T(t)$ is below bottom elevation z_0 of the tank and flow $Q(t)$ is directed outwards. This can be achieved by defining $d(t)$ of equation (4.6) as follows:

$$d(t) = -\bigl(Q(t)|z_0 - H_T(t)| + |Q(t)|(z_0 - H_T(t)) + Q(t)(z_0 - H_T(t))\bigr).$$

Thus, $d(t) < 0$, when both conditions $Q(t) > 0$ and $z_0 - H_T(t) > 0$ are fulfilled. In all other cases $d(t) \geq 0$.

Since the tank has a free water surface, a restriction for the inflow is not applicable. A spillover can be described by the condition $H_T(t) = \min(H_T(t), H_{max})$.

The external flow $Q_s(t)$ given in (4.13) may represent a flow caused by free groundwater of given level $H_g(t)$. Then

$$Q_s(t) = A \frac{H_g(t) - H_T(t)}{T} \tag{4.18}$$

and T can be interpreted as regeneration period. Furthermore, $Q_s(t)$ can be used in order to adapt the simulated water level $H_T(t)$ to the measured one $H_g(t)$. This is helpful during an initial phase, when the initial state of the whole network is unknown. A further discussion of this topic is given in Section 4.3.3.

Pressure tank. Finally, a pressure tank is considered. The total pressure head $H_T(t)$ is given as the sum of bottom elevation z_0, filling height $s(t)$ and air pressure $H_{air}(t)$:

$$H_T(t) = z_0 + s(t) + H_{air}(t). \tag{4.19}$$

Air pressure $H_{air}(t)$ depends on the total height H of the tank and the filling height:

$$H_{air}(s(t)) = \frac{H_0 V_0}{A(H - s(t))}. \tag{4.20}$$

Here, H_0 and V_0 denote initial values for air pressure and air volume in the tank at time t_0. State equation (4.13) is replaced by

$$s'(t) = \frac{1}{A}\bigl(Q_T(t) + Q_s(t)\bigr). \tag{4.21}$$

In order to keep $H_T(t)$ as the state variable, equation (4.21) is modified according to:

$$s'(t) = \frac{d}{dt}\bigl(H_T(t) - z_0 - H_{air}(s(t))\bigr),$$

$$\Rightarrow H_T'(t) = \frac{1}{A}(Q_T(t) + Q_s(t)) + \frac{d}{ds} H_{Air}(s(t)) \cdot s'(t)$$

$$= \frac{1}{A}(Q_T(t) + Q_s(t))\left(1 + \frac{H_0 V_0}{A(H - s(t))^2}\right). \tag{4.22}$$

The filling height in resulting equation (4.22) is now computed from (4.19), (4.20):

$$s(t) = \frac{1}{2}(H_T(t) - z_0 + H) - \sqrt{\left(\frac{1}{2}(H_T(t) - z_0 - H)\right)^2 + \frac{H_0 V_0}{A}}.$$

Coupling conditions stated in (4.14), (4.15), (4.16), (4.17) remain unchanged for a pressure tank.

4.2.3 Energy view

Typically, pumps are the main sources for electric energy consumption within a water supply network. But there may also be other components which require or even produce energy, like UV radiation or generators. Therefore, every element of the mathematical model can be supplied with a specific positve or negative electric power $P_{el}(t)$, which may depend on the current hydraulic state $Q(t)$. When the network element is represented by an edge the current flow $Q(t)$ is given as a state variable. For a tank or pressure tank it is given by $Q_T(t)$, see equation (4.14). On the other hand it is assumend that the energy demand has no influence to the hydraulics. Thus, the total energy consumption of the water work and network operation can be computed subsequent to the hydraulic simulation.

The computation of the electric power within each edge of the network can be done by

$$P_{el}(t) = i_o(t)(\alpha_0 + \alpha_1 Q(t) + \alpha_r Q(t)^r),$$

with coefficients $\alpha_0, \alpha_1, \alpha_r, r$ and switch function $i_o(t) \in \{0, 1\}$. Another possibility is a step function

$$P_{el}(t) = i_o(t) \sum_{i=1}^{n} P_i \, \chi_{[Q_i, Q_{i+1})}(Q(t))$$

with $\chi_{[Q_i, Q_{i+1})}(Q) = \begin{cases} 1 & \text{if } Q_i \leq Q < Q_{i+1} \\ 0 & \text{otherwise} \end{cases}$.

For defining the step function ordered pairs of values (Q_i, P_i), $i = 1, ..., n$ with $Q_1 < Q_2 < ... < Q_n$ and $Q_{n+1} = \infty$ must be given.

By these two approaches of defining electric power consumption many alternatives are covered. When e.g. for a pump $P_{el} = \eta(H_R - H_L)Q$ is assumed, the first approach gives a sufficient approximation by choosing coefficients according to $\alpha_0 = \alpha_r = 0$, $\alpha_1 = \eta \Delta H$. Here, ΔH denotes the nominal delivery height of the pump.

4.3 Simulator TWaveSim

In order to solve the equations presented in the previous section, simulator TWaveSim has been developed. Main purpose of this simulator is an operational calculation of the dynamical hydraulic states within the whole water supply system. At first TWaveSim is supplied with measured data from the last one or two days in order to simulate all states up to the current time. These current states are provided as initial values for the subsequent computation of an optimized operation of the water work and supply network. The operating instructions from the optimization are fed again into the simulator to get a detailed view of future hydraulic and energetic states.

Beside TWaveSim, another simulator Anaconda is in use, see Section 4.4. Since the main purpose of Anaconda is the efficient interplay with the continuous optimization solver, at first TWaveSim is considered more detailed.

4.3.1 Coupling conditions within PDAE system

All modeling equations define a coupled system of partial differential equations (PDEs) and differential-algebraic equations (DAEs), named PDAEs. For their solution first the linear coupling conditions (4.2) are considered. They can easily be defined by incidence matrices M_L and M_R, see [7]. Let the number of nodes and the number of edges in the network be n_N and n_E. Then the $n_N \times n_E$-matrices M_L, M_R are given by matrix elements

$$m_{ij}^L = \begin{cases} 1 & \text{if node } i \text{ is connected to edge } j \text{ via its input} \\ 0 & \text{otherwise} \end{cases},$$

$$m_{ij}^R = \begin{cases} 1 & \text{if node } i \text{ is connected to edge } j \text{ via its output} \\ 0 & \text{otherwise} \end{cases}$$

and the coupling condtions are

$$M_R Q_E^R(t) - M_L Q_E^L(t) + Q_N^S(t) = 0 . \tag{4.23}$$

Here, vectors $Q_E^L(t)$ and $Q_E^R(t)$ denote the left flow respectively right flow variables of all edges and $Q_N^S(t)$ the source or sinks for all nodes.

It is not necessary to define seperate state variables $H_E^L(t)$, $H_E^R(t)$ representing left and right pressure head. Instead, state vector $H_N(t)$ is used, denoting pressure head at each node. Then, $H_E^L(t)$, $H_E^R(t)$ are given by

$$H_E^L(t) = M_L^T H_N(t), \quad H_E^R(t) = M_R^T H_N(t), \tag{4.24}$$

and conditions (4.1) must not be solved explicitely. In order to include nodes representing tanks or pressure tanks, a diagonal $n_N \times n_N$-matrix A_N can be defined with diagonal entries a_{ii}, being the base area A_i of node i. A simple node i without storage capacity is characterized by $a_{ii} = 0$. When equation (4.23) is replaced by

$$A_N H_N'(t) = M_R Q_E^R(t) - M_L Q_E^L(t) + Q_N^S(t), \tag{4.25}$$

equations (4.13) for tanks are included as well. For pressure tanks the right hand side of (4.25) must be complemented according to (4.22). Hence, equations for all nodes are summarized in (4.25). Since matrix A_N is singular, it is a DAE system that will be nonlinear if pressure tanks are included.

This system is supplemented by the nonlinear algebraic equations (4.4), (4.5), (4.15), (4.16), (4.17), (4.8) for connections, valves and pumps and the differential equations (4.6), (4.9) for control valves and control pumps.

4.3.2 Semidiscretization in space and boundary conditions

Solving the partial differential equations for pipes remains to be discussed. System (4.10), (4.11) can be summarized by

$$\partial_t u + M \partial_x u = S(x, u) \tag{4.26}$$

with $u = (H, Q)^T$, $M = \begin{pmatrix} cc0 & \frac{a^2}{gA} \\ gA & 0 \end{pmatrix}$, $S(x, u) = (0, S_f)^T$, $S_f = -\lambda(Q) \frac{Q|Q|}{2DA}$.

The eigenvalues of M are $\lambda_{1,2} = \pm a$. In order to solve this system, the method of lines approach [19] with a conservative finite difference space discretization [16] is applied.

In a first step, space interval $[x_L, x_R]$ is discretized by a constant stepsize Δx leading to $x_L = x_{1/2} < x_{3/2} < \ldots < x_{N+1/2} = x_R$ with cell means $x_i = \frac{1}{2}(x_{i-1/2} + x_{i+1/2})$ for $i = 1, \ldots, N$. For each cell mean, variable $u_i(t) = u(x_i, t)$ is defined and equation (4.26) gives the ODE system

$$\frac{d}{dt} u_i + M \frac{1}{\Delta x}(u_{i+1/2} - u_{i-1/2}) = S(x_i, u_i) . \tag{4.27}$$

Here, $u_{i \pm 1/2}$ approximate values $g(x_{i \pm 1/2}, t)$ of an implicitly defined function $g(x, t)$ given by

$$u(x, t) = \frac{1}{\Delta x} \int_{x - \frac{\Delta x}{2}}^{x + \frac{\Delta x}{2}} g(\xi, t) d\xi . \tag{4.28}$$

Function $g(x, t)$ fullfills

$$\partial_x u(x, t) = \frac{1}{\Delta x} \left(g\left(x + \frac{\Delta x}{2}, t\right) - g\left(x - \frac{\Delta x}{2}, t\right) \right)$$

and

$$\bar{g}_i(t) := \frac{1}{\Delta x} \int_{x_{i-1/2}}^{x_{i+1/2}} g(\xi, t) d\xi = u(x_i, t).$$

Cell means $\bar{g}_i$ coincide with cell centers of u. For that reason well known WENO-schemes from finite volume methods can be applied to compute approximations $g(x_{i \pm 1/2}, t)$ from values $u_i(t)$, see [16].

To get stable discretizations for hyperbolic systems, some upwind information must be included. For this purpose, the so-called Lax-Friedrichs approach is used that splits the flux function into parts with positive and negative eigenvalues, such that

$$M\partial_x u = \frac{1}{2}(M + \lambda^* I)\partial_x u + \frac{1}{2}(M - \lambda^* I)\partial_x u, \tag{4.29}$$

where λ^* denotes the absolute value of the largest eigenvalue, here $\lambda^* = a$. The first summand of (4.29) corresponds to positive eigenvalues and is discretized via an upwind reconstruction $u^-_{i\pm 1/2}$ of $g(x_{i\pm 1/2}, t)$. The second one corresponds to negative eigenvalues and is discretized by a downwind reconstruction $u^+_{i\pm 1/2}$. A first order discretization is given by the choice $u^+_{i+1/2} = u_{i+1}$, $u^-_{i+1/2} = u_i$. To get higher order the third order WENO-method of [12] is applied.

The semi-discretized ODE system (4.27) must be completed by appropriate boundary conditions (BCs). At left and right boundaries $x_{1/2} = x_L$ and $x_{N+1/2} = x_R$, one physical and one numerical BC have to be prescribed for solutions $u_L(t) = u(x_L, t)$ and $u_R(t) = u(x_R, t)$, respectively. Physical BCs are already defined by coupling conditions (4.24), (4.25). To define numerical BCs, friction is neglected, i.e. $S_f = 0$. Thus, from (4.26) it follows:

$$\frac{d}{dt}\underbrace{\left(gAH(x(t), t) - aQ(x(t), t)\right)}_{I_L(t,x)} = 0 \quad \text{for} \quad x(t) = -at + c,$$

$$\frac{d}{dt}\underbrace{\left(gAH(x(t), t) + aQ(x(t), t)\right)}_{I_R(t,x)} = 0 \quad \text{for} \quad x(t) = at + c.$$

Along the characteristic curves $x(t)$ the invariants I_L, I_R can be extrapolated from the interior of $[x_L, x_R]$ to the boundaries x_L, x_R. For the left boundary one gets

$$I_L(t_0, x_1) = I_L\left(t_0 + \frac{x_1 - x_L}{a}, x_L\right), \quad I_L(t_0, x_2) = I_L\left(t_0 + \frac{x_2 - x_L}{a}, x_L\right).$$

Assuming linearity of $I_L(t, x_L)$ in time, for $t_1 = t_0 + \frac{x_1 - x_L}{a}$, $t_2 = t_0 + \frac{x_2 - x_L}{a}$ holds:

$$I_L(t_0, x_L) = I_L(t_1, x_L) - (t_1 - t_0)\frac{I_L(t_2, x_L) - I_L(t_1, x_L)}{t_2 - t_1}$$

$$= I_L(t_1, x_L) - \frac{x_1 - x_L}{a}\frac{I_L(t_2, x_L) - I_L(t_1, x_L)}{\frac{x_2 - x_1}{a}}$$

$$= I_L(t_1, x_L) - \frac{\Delta x}{2}\frac{I_L(t_2, x_L) - I_L(t_1, x_L)}{\Delta x}$$

$$= \frac{3}{2}I_L(t_0, x_1) - \frac{1}{2}I_L(t_0, x_2)$$

For the right BC the calculation is similar. Hence, the two numerical BCs are given by:

$$2(g\,A\,H(x_L,t) - a\,Q(x_L,t)) = 3(g\,A\,H(x_1,t) - a\,Q(x_1,t))$$
$$- (g\,A\,H(x_2,t) - a\,Q(x_2,t)) \qquad (4.30)$$
$$2(g\,A\,H(x_R,t) + a\,Q(x_R,t)) = 3(g\,A\,H(x_N,t) + a\,Q(x_N,t))$$
$$- (g\,A\,H(x_{N-1},t) + a\,Q(x_{N-1},t)) \qquad (4.31)$$

Now, state variables of system (4.27) are collected

$$Y = (H_L, Q_L, \underbrace{H_1, Q_1, ..., H_N, Q_N}_{Y_1}, H_R, Q_R)^T,$$

and equations (4.27), (4.30, 4.31) can be summarized by a semi-explicit DAE-system of type

$$Y_1' = f(t, Y), \qquad (4.32)$$
$$0 = B \cdot Y \qquad (4.33)$$

with a suitable Matrix B.

4.3.3 Initial values

Systems (4.25), (4.32), (4.33) and equations (4.4), (4.5), (4.6), (4.8), (4.9), (4.15), (4.16), (4.17) can be summarized into the semi-explicit DAE system

$$y' = f(t, y, z), \qquad (4.34)$$
$$0 = g(t, y, z). \qquad (4.35)$$

Here, all state variables of all network elements define the vector $\tilde{Y} = (y, z)$. Variables y correspond to ODEs and variables z correspond to algebraic equations. It is assumed that this system is of index one, i.e. $\frac{\partial g}{\partial z}$ is regular.

For the solution of (4.34), (4.35) consistent initial values $(y(t_0), z(t_0))^T = (y_0, z_0)^T$ fulfilling condition $0 = g(t_0, y_0, z_0)$ are required. Since the state of the whole hydraulic network is usually unknown, computation of $(y_0, z_0)^T$ is an important issue. In this context, two approaches are considered in TWaveSim: Warm start and cold start. A warm start can be applied when the model runs routinely in a time loop. In this case, consistent initial values can be determined from the last model run.

In case of a cold start the following procedure is proposed: At initial time t_0 all state variables for flow are set to zero and the pressure variables are set to a constant state H_0: $Q^i(x,t) = 0$, $H^i(x,t) = H_0$, $\forall i$. Obviously, some algebraic equations are not fulfilled and these initial values are not consistent. Therefore, a scale parameter $t_{scale} = \min(\frac{t-t_0}{T}, 1)$ is introduced and the critical algebraic equations (4.2), (4.3), (4.4), (4.8) are replaced by:

$$0 = Q_R^i(t) + Q_R^j(t) - Q_L^k(t) + t_{scale} Q_s(t), \qquad (4.2')$$

$$H_R^i(t) = t_{scale} H_s(t) + (1 - t_{scale}) H_0, \qquad (4.3')$$

$$H_L(t) = H_R(t) + t_{scale} H(Q), \qquad (4.4')$$

$$Q(t) = t_{scale} i(t) \left(\frac{\alpha_0 - d_H(t)}{\alpha_r} \right)^{\frac{1}{r}}. \qquad (4.8')$$

Within time period $t \in [t_0, t_0 + T]$ one gets consistent initial values for the whole network. Nevertheless, the pressure level within the network may be totally wrong. Therefore, at each tank where measurements $h_g(t)$ of the water level are given, external flow $Q_s(t)$ according to (4.18) is applied. After some transition phase the measured and simulated pressure heads at these tanks should agree. When the model is used in the forecast mode, $Q_s(t) = 0$ is assumed.

Another possibility to compute consistent initial values by solving the steady state equations

$$0 = f(t_0, y_0, z_0),$$
$$0 = g(t_0, y_0, z_0),$$

is discussed in [3].

4.3.4 DAE solver

Finally, a DAE solver has to be applied to system (4.34), (4.35). Since TWaveSim is very often executed automatically at the control station of the water work, this solver must be very robust and reliable. On the other hand it should be fast and accurate to allow a coupling to the optimization methods. In this context Rosenbrock methods proved to be good candidates, see [14, 19, 20]. Nevertheless, there are many other efficient classes of solvers which could be used too, see [6].

Within TWaveSim the fourth order Rosenbrock method RODASP [18, 19] is used. It showed good performance in other applications as well, see [6, 13]. One main disadvantage of linearly implicit Rosenbrock methods is the requirement of computing the Jacobian in every timestep. For ODE problems this could be avoided by so-called W-methods [17]. Recently, Jax [8] presented some results to adopt W-methods to DAE problems. Since up to now it was not possible to derive embedded methods of order 4(3), RODASP is still the favoured solver in TWaveSim. Beside the computation of the Jacobian, six evaluations of the right-hand side of the DAE system, one LU decomposition and six back substitutions are required per step. In the context of network applications it is essential to use sparse matrix computations. Iterative solutions of the linear systems did not pay off.

4.4 Simulator Anaconda

Within the project, a second simulation tool named Anaconda, originally developed in [9], is used. The main purpose of this second tool is the possibility to solve continuous nonlinear optimization problems, see Section 5.3. For this optimization, Anaconda must be able to simulate the hydrodynamic network, too. On the other hand, it needs the initial values computed by TWaveSim for starting its computations.

Similar to the approach in TWaveSim, Anaconda also provides the possibility to apply a spatial semidiscretization and use explicit as well as diagonally implicit Runge-Kutta methods for the resulting ODE system. For optimization purposes a so-called first-discretize-then-optimize approach is considered and the underlying adjoint calculus is currently only implemented for one-step methods. Therefore, for a given time discretization $t_0 < t_1 < \cdots < t_N$, we use the implicit box scheme [11] to get a full discretization of the water hammer equations describing the dynamics in the pipes. For consistency reasons, all occuring ordinary differential equations are discretized with the implicit Euler scheme. This way, we finally end up with a fully discretized coupled system of (nonlinear) algebraic equations $E(y,u) = 0$, which depends on state variables

$$y^T = \left(y(t_0)^T, y(t_1)^T, \ldots, y(t_N)^T\right), \tag{4.36}$$

like pressure heads and flow rates, and control variables

$$u^T = \left(u(t_0)^T, u(t_1)^T, \ldots, u(t_N)^T\right), \tag{4.37}$$

e.g. the speed of pumps. Boundary and coupling conditions are already included in $E(y,u)$.

For given initial conditions $y(t_0) = y_0$ and control variables for all timesteps, the simulation task consists of solving these equations. Due to the time-dependent structure and the applied one-step methods, the set of equations $E(y,u)$ can be partitioned,

$$E(y,u) = \begin{pmatrix} y(t_0) - y_0 \\ F(t_0, t_1, y(t_0), y(t_1), u(t_0), u(t_1)) \\ \vdots \\ F(t_{N-1}, t_N, y(t_{N-1}), y(t_N), u(t_{N-1}), u(t_N)) \end{pmatrix} = 0, \tag{4.38}$$

and solved for $y(t_j)$ time step by time step ($j = 1, \ldots, N$). Therefore, we apply Newton's method to the subsets

$$F(t_{j-1}, t_j, y(t_{j-1}), y(t_j), u(t_{j-1}), u(t_j)) = 0 \tag{4.39}$$

of E. Here, we can exploit the sparsity structure of the underlying Jacobian matrix by using an appropriate solver for the sets of linear equations [2]. Unfortunately, the discretized model equations of water supply networks do not always yield unique solutions. We treat the underlying singularities as described in [10].

4.5 Test application

4.5.1 Problem setup

In this section a test scenario is presented. This example includes all network elements from Table 4.1. But in order to specify explicitly all equations, it can not be too complex. Figure 4.1 shows the structure of the problem.

A water work is supplied by two water sources, delivering flows $Q_{in}^{(1)}(t)$ and $Q_{in}^{(2)}(t)$ into node 1 and tank 4. Node 1 is connected to tank 2, and from tank 2 water flow to tank 4 is controlled by valve 3. From tank 4 water is pumped into the pressure tank 6. This pressure tank provides the pressure zone with water. At nodes 8, 12 and tank 10 water is taken out. Node 8 is connected to tank 6 by a pipe and from that node water is pumped into tank 10. Node 12 is linked to the pressure tank by a connection and to tank 10 by a backflow preventer. Energy demand is not considered in this example.

The state vector of the whole system is given by

$$y = (H_1, Q_1, H_2, Q_3, s_3, H_4, Q_5, H_6, Q_{7L}, H_{71}, Q_{71}, H_{72}, Q_{72}, Q_{7R},$$
$$H_8, Q_9, H_{10}, Q_{11}, H_{12}, Q_{13}, s_{13})^T .$$

All variables are time dependend. In order to simplify the system, the pipe is resolved by two cells only and discretized in space by the first order approach. This would not be the case in a realistic application.

The 21 modeling equations according to Sections 4.2.1, 4.2.2 are presented below. The DAE system consists of 10 ODEs and 11 algebraic equations.

$$0 = Q_{in}^{(1)}(t) - Q_1(t) \tag{4.40}$$
$$0 = H_1(t) - H_2(t) - H^1(Q_1(t)) \tag{4.41}$$
$$A_2 H_2'(t) = Q_1(t) - Q_3(t) + Q_{2s}(t) \tag{4.42}$$
$$0 = s_3^2(t)(H_2(t) - H_4(t)) - \tilde{\zeta}_3 |Q_3(t)| Q_3(t) \tag{4.43}$$

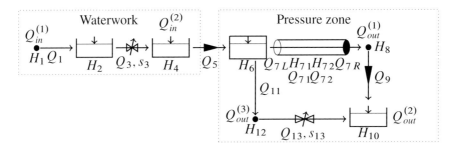

Figure 4.1. Network and state variables for test case

4 Hydraulic modeling and energy view

$$s'_3(t) = \frac{H_4^t - H_4(t)}{\alpha_3}\left(f^+(1-s_3(t)) + f^-(s_3(t)-0)\right) \tag{4.44}$$

with $f^+ = \dfrac{\text{sgn}(H_4^t - H_4(t)) + 1}{2}$, $f^- = \dfrac{1 - \text{sgn}(H_4^t - H_4(t))}{2}$

$$A_4 H'_4(t) = Q_3(t) - Q_5(t) + Q_{in}^{(2)} + Q_{4s}(t) \tag{4.45}$$

$$0 = Q_5(t) - \left(\frac{\alpha_{05} - H_6(t) + H_4(t)}{\alpha_{r5}}\right)^{\frac{1}{r_5}} \tag{4.46}$$

$$A_6 H'_6(t) = (Q_5(t) - Q_{7L}(t) - Q_{11}(t) + Q_{6s}(t))\left(1 + \frac{H_{06} V_{06}}{A_6(H^6 - s(t))^2}\right) \tag{4.47}$$

with $s(t) = \dfrac{1}{2}(H_6(t) - z_{06} + H^6) - \sqrt{\left(\dfrac{1}{2}(H_6(t) - z_{06} - H^6)\right)^2 + \dfrac{H_{06} V_{06}}{A_6}}$

$$0 = 2(gA_7 H_6(t) - a Q_{7L}(t)) - 3(gA_7 H_{71}(t) - a Q_{71}(t)) + (gA_7 H_{72}(t) - a Q_{72}(t)) \tag{4.48}$$

$$H'_{71}(t) = -\frac{a^2}{g A_7}\frac{Q_{3/2} - Q_{1/2}}{\Delta x} + a\frac{H_{3/2} - H_{1/2}}{\Delta x} \tag{4.49}$$

with

$Q_{1/2} = Q_{7L}$, $H_{1/2} = H_6$, $Q_{3/2} = \dfrac{1}{2}(Q_{71} + Q_{72})$, $H_{3/2} = \dfrac{1}{2}(H_{71} + H_{72})$

$$Q'_{71}(t) = -g A_7 \frac{H_{3/2} - H_{1/2}}{\Delta x} + a\frac{Q_{3/2} - Q_{1/2}}{\Delta x} - \lambda(Q_{71}(t))\frac{Q_{71}(t)|Q_{71}(t)|}{2 D_7 A_7} \tag{4.50}$$

with $\lambda(Q)$ according to 4.12

$$H'_{72}(t) = -\frac{a^2}{g A_7}\frac{Q_{5/2} - Q_{3/2}}{\Delta x} + a\frac{H_{5/2} - H_{3/2}}{\Delta x} \tag{4.51}$$

with $Q_{5/2} = Q_{7R}$, $H_{5/2} = H_8$

$$Q'_{72}(t) = -g A_7 \frac{H_{5/2} - H_{3/2}}{\Delta x} + a\frac{Q_{5/2} - Q_{3/2}}{\Delta x} - \lambda(Q_{72}(t))\frac{Q_{72}(t)|Q_{72}(t)|}{2 D_7 A_7} \tag{4.52}$$

$$0 = 2(gA_7 H_8(t) + a Q_{7R}(t)) - 3(gA_7 H_{72}(t) + a Q_{72}(t)) + (gA_7 H_{71}(t) + a Q_{71}(t)) \tag{4.53}$$

$$0 = Q_{7R}(t) - Q_9(t) - Q_{out}^{(1)}(t) \tag{4.54}$$

$$0 = Q_9(t) - i_{o9}(t)\left(\frac{\alpha_{09} - H_{10}(t) + H_8(t)}{\alpha_{r9}}\right)^{\frac{1}{r_9}} \tag{4.55}$$

$$A_{10} H'_{10}(t) = Q_9(t) + Q_{13}(t) - Q_{out}^{(2)} + Q_{10s}(t) \tag{4.56}$$
$$0 = H_6(t) - H_{12}(t) - H^{11}(Q_{11}(t)) \tag{4.57}$$
$$0 = Q_{11}(t) - Q_{13}(t) - Q_{out}^{(3)}(t) \tag{4.58}$$
$$0 = s_{13}^2(t)(H_{12}(t) - H_{10}(t)) - \tilde{\zeta}_{13}|Q_{13}(t)|Q_{13}(t) \tag{4.59}$$
$$s'_{13}(t) = \frac{Q_{13}(t)}{\alpha_{13}} \left(f^+(1 - s_{13}(t)) + f^-(s_{13}(t) - 0) \right) \tag{4.60}$$
$$\text{with } f^+ = \frac{\text{sgn}(Q_{13}(t)) + 1}{2}, \quad f^- = \frac{1 - \text{sgn}(Q_{13}(t))}{2}$$

To simulate the system, all constants, inflow and outflow functions and initial conditions must be specified. In- and outflow functions are choosen as:

$$Q_{in}^{(1)}(t) = 0.25 + 0.1 \sin\left(\frac{2\pi t}{7200}\right), \quad Q_{in}^{(2)}(t) = 0.25$$

$$Q_{out}^{(1)}(t) = 0.3, \quad Q_{out}^{(2)}(t) = 0.05 + 0.025 \cos\left(\frac{2\pi t}{3600}\right), \quad Q_{out}^{(3)}(t) = 0.15.$$

The units are not stated here, they are assumed to be in the SI system. Pressure losses of the two connections are $H^1(Q) = 2$, $H^{11}(Q) = 5$ and the valve coefficients are $\tilde{\zeta}_3 = \tilde{\zeta}_{13} = 1$, $\alpha_3 = 100$, $\alpha_{13} = 1$. Target value of control valve 3 is pressure head $H_4^t = 35$. The two pumps are defined by coefficients $\alpha_{05} = 100$, $\alpha_{r5} = -300$, $r_5 = 2.5$, $\alpha_{09} = 68$, $\alpha_{r5} = -10000$, $r_9 = 2$. Pipe 5 is running permanently, whereas pipe 9 is switched off after 18 hours and switched on again after 48 hours:

$$i_{o9}(t) = \begin{cases} 0 & \text{if } t \in [64800, 172800] \\ 1 & \text{otherwise} \end{cases}.$$

Parameters of the pipe are choosen as $g = 9.81$, $a = 1414$, $D_7 = 0.5$, $A_7 = 0.0625\pi$, $\Delta x = 1000$, $k = 0.0005$, $\nu = 1.31 \cdot 10^{-6}$. Finally, the coefficients of the three tanks and the pressure tank are given by $A_2 = 800$, $A_4 = 300$, $A_6 = 400$, $A_{10} = 200$, $z_{06} = 80$, $H^6 = 10$, $H_{06} = 10$, $V_{06} = 500$.

As stated in Section 4.3.3 at initial time $t_0 = 0$ all initial values for $Q_i(t_0)$ are set to zero and $H_i(t_0)$ are set to $H_0 = 102.5$. The time period for getting consistent initial values is set to $T = 10$.

The total simulation time is three days, i.e. $t \in [0, 259200]$. To get correct initial states, artificial flow terms $Q_s(t)$ in each storage element are applied during the first eight hours:

$$Q_{2s}(t) = \frac{45 - h_2(t)}{100}, \quad Q_{4s}(t) = \frac{35 - h_4(t)}{100} \tag{4.61}$$
$$Q_{6s}(t) = \frac{82 - h_6(t)}{100}, \quad Q_{10s}(t) = \frac{110 - h_{10}(t)}{100}. \tag{4.62}$$

Spillover or emtying of tanks is not considered.

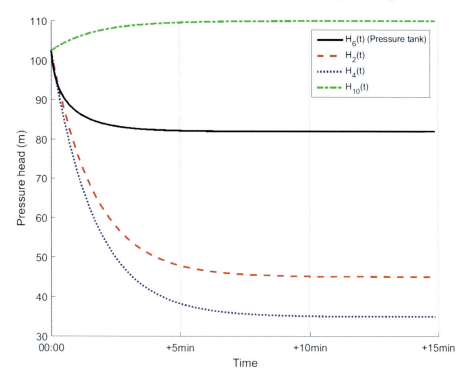

Figure 4.2. Pressure heads of tanks during initial phase

4.5.2 Simulation results

First, TWaveSim has been applied to simulate the test scenario. The absolute and relative tolerances of integrator RODASP were set to 10^{-6}. In total, 2347 successful time steps, 817 rejected time steps and 34,326 function evaluations of the right hand side of the DAE system were necessary. Since numerical accuracy and efficiency is not the main topic here, only selected simulation results are shown exemplary.

In Figure 4.2 the initial transient phase is shown during the first 15 minutes. Due to the artficial flow terms (4.61), (4.62) the pressure heads in the three tanks and the pressure tank starting from common initial values reach their defined target values of 45, 35, 82 and 110 m within about 10 minutes.

Figure 4.3 shows pressure heads for tanks 2 and 4 over the whole time interval. Additionally, the opening degree of control valve 3 is presented. It can be seen that the control valve is able to induce a constant pressure head at tank 4 by varying its opening degree. The small oscillations in pressure head of tank 2 and opening degree of valve 3 are caused by the sinusoidal input flow $Q_{in}^{(1)}(t)$.

Finally, in Figure 4.4 flow through pump 9 and valve 13 are shown. The pump does not operate in time interval between 18 and 48 hours. Therefore approximately after 40 hours, pressure head of tank 10 sinks below the one of node 12 and valve 13

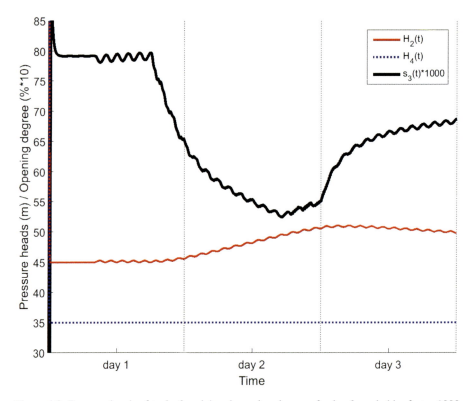

Figure 4.3. Pressure heads of tanks 2 and 4 and opening degree of valve 3, scaled by factor 1000

opens suddenly at that time. When pump 9 operates again, pressure head of tank 10 is above that of node 12 and valve 13 acts as a backflow preventer again, i.e. it closes. At the switching times of valve 13 strong pressure peaks can be observed. They are caused by the sudden opening and closing of the valve. Moreover, the oscillating bahaviour of flow through that valve is caused by the sinusoidal exctraction term $Q_{out}^{(2)}(t)$.

Next, simulator Anaconda has been applied to the same example. Since Anaconda is not able to deal with artificial sources (4.61), (4.62), it must be provided with consistent inital values. These values were taken from the results of TWaveSim for time $t = 28{,}800$. Thus, Anaconda simulates only the time period from 8 to 72 hours. The differences between the results of TWaveSim and Anaconda are negligible from a practical point of view. As an example, Figure 4.5 shows a detailed view. It can been seen, that Anaconda uses much larger time steps than TWaveSim. Since additionally the order of the time stepping of Anaconda is only one compared to order 4 of TWaveSim, the solution is smoothed very much. The small oscillations in the result of TWaveSim result from the regulation of the valve in combination with the sinusoidal behavior of the inflow $Q_{in}^{(1)}(t)$.

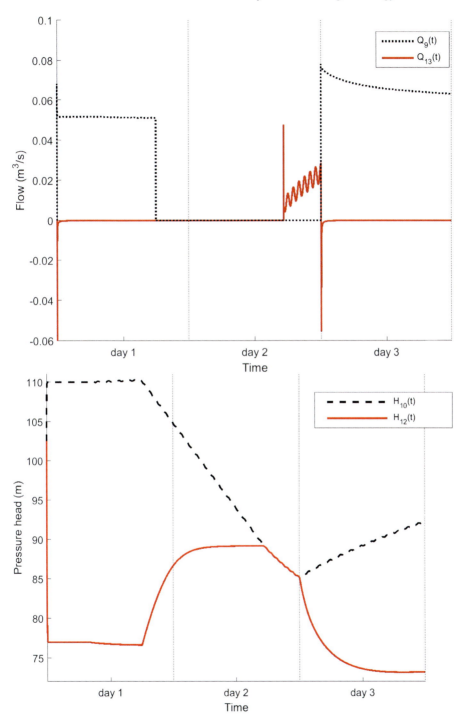

Figure 4.4. Flow through pump 9 and valve 13 and pressure heads of tank 10 and node 12

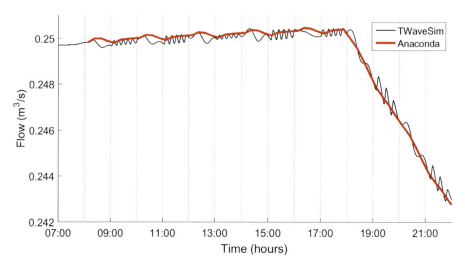

Figure 4.5. Comparison of results for flow through valve 3 from TWaveSim and Anaconda

4.6 Conclusion

With the presented network approach it is possible to model the most important processes in the chain of water extraction, water treatment and water distribution. The idea is to identify important network components and describe them mathematically. Coupling of these elements creates a mathematical model of the drinking water system. Unlike many commercial models, a fully dynamic modelling approach has been choosen. This transient simulation of the hydrodynamics allows at the same time an accurate determination of the electrical energy requirement of all network components. This is the prerequisite for the subsequent optimization for energy optimal control of the drinking water system.

With TWaveSim and Anaconda two simulation methods were developed. The first method allows an accurate simulation and the calculation of consistent states of the network, while the second method can be used simultaneously for continuous optimization in addition to the simulation. The suitability of the model approach and the differences of the two simulators were demonstrated by an extensive test example.

Bibliography

[1] J. Abreu, E. Cabrera, J. Izquierdo and J. Garcia-Serra, Flow Modeling in Pressurized Systems Revisited. *Journal of Hydraulic Engineering* 125(11) (1999), 1154–1169. http://link.aip.org/link/?QHY/125/1154/1.

[2] T. A. Davis, title = Direct Methods for Sparse Linear Systems (Fundamentals of Algorithms 2), Society for Industrial and Applied Mathematics, Philadelphia, PA, USA, 2006.

[3] D. M. Dreistadt, Modellbildung und Berechnung konsistenter Anfangszustände für dynamische Simulationen in der Wasserversorgung, MA thesis. Hochschule Bonn-Rhein-Sieg, FB EMT, 2015.

[4] M. Günther, A PDAE Model for Interconnected Linear RLC Networks, *Mathematical and Computer Modelling of Dynamical Systems* 7(2) (2001), 189–203.

[5] C. Hähnlein, Numerische Modellierung zur Betriebsoptimierung von Wasserverteilnetzen, PhD thesis. Technische Universität Darmstadt, 2008.

[6] E. Hairer and G. Wanner, *Solving Ordinary Differential Equations II*, Springer, Berlin, Heidelberg, 1996. http://dx.doi.org/10.1007/978-3-642-05221-7.

[7] C. Huck, L. Jansen, and C. Tischendorf, A Topology Based Discretization of PDAEs Describing Water Transportation Networks. *PAMM* 14(1) (2014), 923–924. http://dx.doi.org/10.1002/pamm.201410442.

[8] T. Jax and G. Steinebach, Generalized ROW-type methods for solving semi-explicit DAEs of index-1. *Journal of Computational and Applied Mathematics* 316 (2017), 213–228. http://www.sciencedirect.com/science/article/pii/S0377042716303934.

[9] O. Kolb, *Simulation and Optimization of Gas and Water Supply Networks*, Verlag Dr. Hut, München, 2011.

[10] O. Kolb, P. Domschke and J. Lang, Modified QR decomposition to avoid non-uniqueness in water supply networks with extension to adjoint calculus. *Procedia Computer Science* 1(1) (2010), 1421–1428.

[11] O. Kolb, J. Lang and P. Bales, An implicit box scheme for subsonic compressible flow with dissipative source term. *Numerical Algorithms* 53(2) (2010), 293–307.

[12] A. Kurganov and D. Levy, A Third-Order Semidiscrete Central Scheme for Conservation Laws and Convection-Diffusion Equations. *SIAM Journal on Scientific Computing* 22(4) (2000), 1461–1488. http://dx.doi.org/10.1137/S1064827599360236.

[13] J. Rang, High order time discretisation methods for incompressible Navier-Stokes equations. *PAMM* 16(1) (2016), 759–760. http://dx.doi.org/10.1002/pamm.201610368.

[14] P. Rentrop and G. Steinebach, Model and numerical techniques for the alarm system of river Rhine. *Surveys Math. Industry* 6 (1997), 245–265.

[15] L. A. Rossman, *EPANET 2 users manual*, U.S. Environmental Protection Agency, Cincinnati, OH, 2000.

[16] C.-W. Shu, High Order Weighted Essentially Nonoscillatory Schemes for Convection Dominated Problems. *SIAM Review* 51(1) (2009), 82–126. http://dx.doi.org/10.1137/070679065.

[17] T. Steihaug and A. Wolfbrandt, An attempt to avoid exact Jacobian and nonlinear equations in the numerical solution of stiff differential equations. *Mathematics of Computation* 33(146) (1979), 521–534.

[18] G. Steinebach, *Order reduction of ROW methods for DAEs and method of lines applications*, Preprint. Technische Hochschule Darmstadt, Fachbereich Mathematik, 1995. https://books.google.de/books?id=6TRDHQAACAAJ.

[19] G. Steinebach and P. Rentrop, An Adaptive Method of Lines Approach for Modeling Flow Transport in Rivers. In *Adaptive Method of Lines*, Chapman and Hall/CRC, 2001. http://dx.doi.org/10.1201/9781420035612.ch6.

[20] G. Steinebach, R. Rosen, and A. Sohr, Modeling and Numerical Simulation of Pipe Flow Problems in Water Supply Systems. In *Mathematical Optimization of Water Networks*, Springer Nature, 2012, 3–15. http://dx.doi.org/10.1007/978-3-0348-0436-3_1.

Chapter 5
Optimization

Björn Geißler, Alexander Martin, Antonio Morsi, Maximilian Walther, Oliver Kolb, Jens Lang, and Lisa Wagner

Abstract. The core component of EWave is constituted by a receding horizon optimal control algorithm that is implemented in the EWave optimization module (EWave-OPT). EWave-OPT is made up of two major components, the discrete optimization module, which is responsible for computing optimal discrete switching decisions on the basis of a quasi-stationary approximation of the physical reality, and the continuous optimization module (EWave-NOPT) that, in a second step, computes optimal values for the continuous control variables subject to fixed discrete controls and an instationary, highly accurate pipeflow model.

5.1 Introduction

The EWave optimization module (EWave-OPT) is responsible for computing an optimal control (e.g. one that leads to minimum overall energy consumption) for a network over a given time horizon.

EWave-OPT is composed of two major components, the discrete optimization module (EWave-DOPT), and the continuous optimization module (EWave-NOPT).

EWave-OPT is called periodically with scenarios following a receding horizon. Before each call, a simulation is performed to predict the state of the network at the beginning of the time horizon from the latest measured data points. This is necessary due to the fact that it takes some time to solve the optimization problem and to apply the computed control.

On each call of EWave-OPT, a solution to the combinatorial decision variables is computed first by calling EWave-DOPT. This is done by applying a quasi-stationary, but sufficiently detailed approximation of the instationary physical model. The applied approximation is in particular based on a piecewise linear approximation of the nonlinear constraints, see Section 5.2. In a second step, the computed combinatorial decisions are used to setup and solve a continuous nonlinear optimization problem (NLP), see Section 5.3 during a call to EWave-NOPT. While the combinatorial decisions might not change during this second step, optimal values for the continuous control variables are computed on the basis of the highly accurate instationary physical model used by EWave-NOPT.

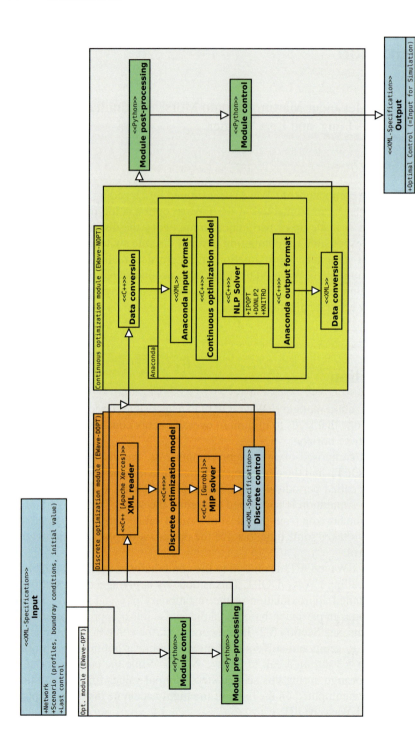

Figure 5.1. Schematic view of EWave-OPT

This two-step approach is necessary, since solving the involved optimization models containing both nonlinear and nonconvex constraints and combinatorial decision variables in acceptable computing time is only possible for very small networks and few timesteps with today's algorithms and software. Solving such an optimization model for a real-world network and a over a time-horizon of several hours in an one-step approach is impractical today. Besides the vast number of decision variables, the main reason why such kind of optimization model is hard to solve, is the non-convexity of its set of feasible solutions, which in turn is mainly due to the presence of integrality restrictions and nonlinear equality constraints. The latter arise naturally for instance to describe the pressure drop along a pipe.

Data exchange between EWave-DOPT and EWave-NOPT is realized via a tailored XML data format that is also used for the input and output of EWave-OPT.

Figure 5.1 contains a schematic overview of EWave-OPT, the data flow, as well as the programming languages and the most important external libraries used for implementing the algorithms.

The XML input data is composed of a description of the network, including its topology and a detailed physical and technical specification of all of its elements, scenario data, i.e., (demand) profiles, boundary conditions, like, e.g., minimum pressure requirements, an intial state containing for instance the filling levels of all tanks, and the network control for the time before the beginning of the optimization horizon. The latter is for instance needed to respect minimum up- or downtime constraints for pumps during optimization.

EWave-OPT, which is implemented in Python is then responsible for receiving this data, passing it to EWave-DOPT and EWave-NOPT, calling EWave-DOPT and EWave-NOPT in an appropriate way, and finally for passing the computed optimal discrete-continuous control as an XML-file back to the caller.

Both EWave-DOPT and EWave-NOPT are implemented in C++ and make use of the Apache Xerces library for reading and writing XML data. EWave-DOPT internally uses the optimization software Gurobi to deal with the linear mixed-integer optimization problems (MIPs) that have to be solved during the algorithm descibed in Section 5.2. EWave-NOPT makes use of the software package Anaconda for solving continuous optimization problems on networks, which in turn make use of Ipopt, Donlp2, and Knitro to solve nonlinear optizaton problems (NLPs), see Section 5.3.

5.2 Discrete optimization

5.2.1 Water supply network model

The simplified hydraulic model used to obtain a finite dimensional mixed-integer nonlinear programming model (MINLP) is based on the very accurate instationary model presented in Section 4.2. For further references to the presented hydraulic equations we refer to the reference therein.

The base simplification arises from the assumption that a quasi-stationary model is sufficiently accurate for optimizing discrete controls (mainly pump on-off schedules) at an hourly up to quarter-hourly resolution. Tanks are the only elements which can store water, for all further components storage effects are neglected. We remark that a stationarity assumption is not necessary, but it reduces the model size and complexity in contrast to applying an appropriate numerical discretization scheme.

To obtain a model, which can be handled efficiently by MINLP solution approaches further simplification are applied.

We start with a description of the underlying network topology, continue with the temporal and spatial discretization, and give a short introduction to the notation that is used to present our MINLP model. Subsequentely, we introduce our model variables and constraints that are added for all nodes and edges, and on top of that for the specific components junctions, tanks, pipes, pumps, valves, and connections. We conclude the description with a compact summary of our entire MINLP model.

Network topology. We model a water supply network as a directed graph $G = (V, A)$. If water flows from i to j on an edge $a = (i, j)$, the sign of the flow is positive and if water flows from j to i, the flow along a has a negative sign. The set of edges ending at a node $i \in V$ is denoted by $\delta^{\text{in}}(i)$ and the set of edges starting at i is denoted by $\delta^{\text{out}}(i)$.

The set of nodes V consists of the set of junctions V_{ju} and the set of tanks V_{ta}, i.e., $V = V_{ju} \cup V_{ta}$.

Further, the set of edges A is composed of the set of pipes A_{pi}, the set of valves A_{va}, the set of pumps A_{pu}, and the set of connections A_{co}, i.e., $A = A_{pi} \cup A_{va} \cup A_{pu} \cup A_{co}$.

In more detail, the set of valves A_{va} contains the set of gate valves A_{gv}, the set of check valves A_{xv}, and the set of control valves A_{cv}, i.e., $A_{va} = A_{gv} \cup A_{xv} \cup A_{cv}$.

Moreover, we have $A_{pu} = A_{pv} \cup A_{pf}$, where A_{pv} denotes the set of pumps with variable speed and A_{pf} denotes the set of pumps with a fixed speed.

Finally, for connections, we distinguish A_{co} between hydraulic coupled connections with given pressure loss curve A_{ch} and hyraulic decoupled connections with given inlet pressure A_{cn}.

Discretization. We use a quasi-stationary hydraulic model with a predefined discretization of the considered timespan into N intervals $K := \{1, \ldots, N\}$. For convenience we assume equidistant intervals of length Δt. In our model we assume for a variable y defined on some node or edge that its value does not change within a temporal discretization interval, i.e., $y(t_1, x) = y(t_2, x)$ for all $t_1, t_2 \in [(k-1)\Delta t, k\Delta t)$, $k \in K$.

In the spatial domain we choose a two-endpoint discretization, namely at the beginning ($x = 0$) and at the end ($x = L_a$) for each edge $a \in A$ with length L_a.

Notation. Throughout this section we will introduce a number of variables that are associated with either a node $i \in V$ or with the beginning and end of an edge $a \in A$, respectively. For a variable named y, we write $y^{i,k}$ if it is associated with a node $i \in$

V and time interval $k \in K$. We write $y_L^{a,k}$ or $y_R^{a,k}$ if it is associated with the beginning or end of an edge $a \in A$ and time interval $k \in K$, respectively. If the value of y is constant along an edge, i.e., $y_L^{a,k} = y_R^{a,k}$, we simply write $y^{a,k}$. This may be the case due to edges $a \in A \setminus A_{pi}$ with length $L_a = 0$ or an implication of the steady state assumption.

During the derivation of individual component models we sometimes omit the respective element or time indices, when they are clear from the context.

The lower and upper bound of a variable y is denoted by $[y]^-$ and $[y]^+$, respectively.

To avoid confusing conversion factors all model constants, variables, and equations are introduced in SI units unless otherwise stated, although different units or apprioriate scalings are essential to obtain a numerically robust model.

Nodes. For each node $i \in V$ we are given its geodetic height z_{0i} in m, which is related to the pressure head $h^{i,k}$ via the equation

$$h^{i,k} = z_{0i} + \frac{p^{i,k}}{g\rho_0}, \qquad (5.1)$$

where p denotes the pressure in Pa, $g \approx 9.80665\,\text{m/s}^2$ is the acceleration due to gravity and ρ_0 a constant reference density of water, e.g., the density of water under normal conditions $\rho_0 = 1000\,\text{kg/m}^3$.

In our model, we introduce the variables

$$h^{i,k} \geq z_0, \qquad \forall i \in V \ \forall k \in K, \qquad (5.2\text{a})$$
$$h^{i,k} \in \mathbb{R}, \qquad \forall i \in V \ \forall k \in K, \qquad (5.2\text{b})$$

for the pressure head at node i, the variables

$$q_{ext}^{i,k} \in \mathbb{R}, \qquad \forall i \in V \ \forall k \in K, \qquad (5.2\text{c})$$

for the volumetric rate of external flow supplied to the network at node i, and the variables

$$h_{ext}^{i,k} \in \mathbb{R}, \qquad \forall i \in V \ \forall k \in K, \qquad (5.2\text{d})$$

for the head of an external source that supplies flow to the network at node i. The pressure head are given in m and volumetric flow rates are given in m^3/s.

Typically, either the external supply or the pressure head is fixed at a node, i.e., exactly one of the variables $h_{ext}^{i,k}$ or $q_{ext}^{i,k}$ is fixed, while the other remains free. This is reflected by setting the respective lower and bounds of the variable to an equal value and to appropriate large value, respectively. Nonetheless this is not mandatory and it may be possible to optimize the respective values, such as external supplies from wells or vendors.

Edges. For each edge $a = (i, j) \in A$ we introduce variables

$$q_L^{a,k}, q_R^{a,k} \in \mathbb{R}, \quad \forall a \in A \; \forall k \in K, \tag{5.3a}$$

for the volumetric flow rate in m³/s at the beginning and at the end point of a. We also introduce variables

$$h_L^{a,k}, h_R^{a,k} \in \mathbb{R}, \quad \forall a \in A \; \forall k \in K, \tag{5.3b}$$

for the pressure head in m at the beginning and at the end point of a.

Junctions. For a junction $i \in V_{ju} \subseteq V$ no further variables are necessary in addition to those that have already been introduced for all nodes.

But we introduce the flow conservation constraints

$$q_{ext}^{i,k} = \sum_{a \in \delta^{out}(i)} q_L^{a,k} - \sum_{a \in \delta^{in}(i)} q_R^{a,k}, \quad \forall i \in V \; \forall k \in K. \tag{5.4a}$$

We also add the pressure head propagation constraints

$$h^{i,k} = h_L^{a,k}, \quad \forall a \in \delta^{out}(i) \; \forall k \in K, \tag{5.4b}$$

$$h^{i,k} = h_R^{a,k}, \quad \forall a \in \delta^{in}(i) \; \forall k \in K. \tag{5.4c}$$

Tanks. For each tank $i \in V_{ta} \subseteq V$ we are additionally given its cross-sectional area A_i in m², its height H_i in m, a friction coefficient ζ_i in s²/m⁵, and its initial filling level $s^{i,0}$ in m immediately before the optimization horizon.

In addition to the variables introduced in Section 5.2.1, we add a variable $s^{i,k}$ for $k \in K$ for the filling level of tank i at time k to our model. Further, we add the constraints

$$h^{i,k} = z_0 + s^{i,k}, \quad \forall i \in V_{ta} \; \forall k \in K, \tag{5.5a}$$

$$s^{i,k} \in [0, H_i] \subset \mathbb{R}, \quad \forall i \in V_{ta} \; \forall k \in K \tag{5.5b}$$

that relate the pressure head to the filling level of the tank. The change in pressure head due to fittings at the inlet and outlet of a tank are modeled by the constraints

$$h^{i,k} - h_L^{a,k} = \zeta_i |q_L^{a,k}| q_L^{a,k}, \quad \forall i \in V_{ta} \; \forall a \in \delta^{out}(i) \; \forall k \in K, \tag{5.5c}$$

$$h_L^{a,k} - h^{i,k} = \zeta_i |q_R^{a,k}| q_R^{a,k}, \quad \forall i \in V_{ta} \; \forall a \in \delta^{in}(i) \; \forall k \in K, \tag{5.5d}$$

$$h_{ext}^{i,k} - h^{i,k} = \zeta_i |q_{ext}^{i,k}| q_{ext}^{i,k}, \quad \forall i \in V_{ta} \; \forall k \in K. \tag{5.5e}$$

To model the change of filling level over time, we also introduce the constraints

$$\frac{A_i}{\Delta t}(s^{i,k-1} - s^{i,k}) = \sum_{a \in \delta^{out}(i)} q_L^{a,k} - \sum_{a \in \delta^{in}(i)} q_R^{a,k} - q_{ext}^{i,k}, \quad \forall i \in V_{ta} \; \forall k \in K. \tag{5.5f}$$

Pipes. For a pipe $a = (i, j) \in A_{pi} \subseteq A$, we are given its length L_a its diameter D_a and roughness ζ_a in m. Our model of pipeline hydraulics is based on the so-called water hammer equations

$$\frac{\partial h}{\partial t} + \frac{c^2}{gA}\frac{\partial q}{\partial x} = 0, \tag{5.6}$$

$$\frac{\partial q}{\partial t} + gA\frac{\partial h}{\partial x} = -\lambda(q)\frac{q|q|}{2DA}. \tag{5.7}$$

Equation (5.6) is the continuity equation describing conservation of mass and Equation (5.7) describes conservation of momentum. In the stationary case all time derivatives vanish such that Equation (5.6) simply states that flow is constant along a pipe, which justifies our decision to introduce only a single variable for the volumetric flow rate along a pipe for each point in time. Setting $\partial q/\partial t = 0$ in Equation (5.7) and then integrating both sides of the equation along the pipe yields the stationary momentum equation

$$h(0) - h(L) = \frac{\lambda(q)}{2gDA^2}|q|q. \tag{5.8}$$

Thus, we add the constraints

$$q_L^{a,k} = q_R^{a,k} = q^{a,k} \tag{5.9a}$$

$$h_L^{a,k} - h_R^{a,k} = \phi^{a,k}, \quad \forall a \in A_{pi} \; \forall k \in K \tag{5.9b}$$

$$\phi^{a,k} = \lambda_a(q^{a,k})\frac{L_a}{2gD_aA_a^2}|q^{a,k}|q^{a,k}, \quad \forall a \in A_{pi} \; \forall k \in K \tag{5.9c}$$

$$\phi^{a,k} \in \mathbb{R}, \quad \forall a \in A_{pi} \; \forall k \in K \tag{5.9d}$$

to our model, where λ_a denotes the friction factor. We use Nikuradse's Equation [21, 20] for computing the friction factor, which does not depend on q:

$$\lambda_a = \frac{1}{\left(2\log_{10}\left(\frac{\zeta_a}{3.71D_a}\right)\right)^2}. \tag{5.10}$$

We remark that other formulas for the friction factor, such as the Swamee-Jain approximation [30] (see Chapter 4) or the implicit Colebrook-White formula [4] could be used with our solution approach.

Valves. We distinuish between gate valves that allow bidirectional flow, check valves that are used to prevent reverse flow and control valves that are used to reduce the flow rate or the pressure head. For each type of valve we provide two models. One model is used for automated valves, whose control is subject to optimization. The other model must be employed for valves that are not remotely controlable or whose control should not be optimized for some other reason, e.g., during periods of maintenance work.

For convenience we add the variables $\Delta h^{a,k}$ for the change in pressure head to our model for $a \in A_{va}$ and for $k \in K$. We also add the defining constraints

$$q_L^{a,k} = q_R^{a,k} \tag{5.11a}$$

$$\Delta h^{a,k} = h_L^{a,k} - h_R^{a,k}, \quad \forall a \in A_{va} \ \forall k \in K, \tag{5.11b}$$

$$\Delta h^{a,k} \in \mathbb{R}, \quad \forall a \in A_{va} \ \forall k \in K. \tag{5.11c}$$

Since valves are considered to have a very short length, i.e., $L_a = \epsilon > 0$, we added Equation (5.11a).

Gate valves. For a gate valve $a \in A_{gv} \subseteq A_{va} \subseteq A$, we are given the area of its inlet A_a, when it is fully open and its friction coefficient ζ_a. To model a gate valve, we have to guarantee that the equations

$$\left(u^{a,k}\right)^2 \Delta h^{a,k} = \phi^{a,k}, \quad \forall a \in A_{gv} \ \forall k \in K, \tag{5.12}$$

with

$$\phi^{a,k} = \frac{\zeta_a}{2gA_a^2} |q_L^{a,k}| q_L^{a,k}, \quad \forall a \in A_{gv} \ \forall k \in K, \tag{5.13a}$$

$$\phi^{a,k} \in \mathbb{R}, \quad \forall a \in A_{gv} \ \forall k \in K, \tag{5.13b}$$

are satisfied. Here $u^{a,k} \in [0,1]$ denotes the opening degree of a at time k.

Let $K_{opt}(a) \subseteq K$ be such that the control of a is subject to optimization during the timespans $[k, k+1)$ for $k \in K_{opt}(a)$ and let $K_{sim}(a) := K \setminus K_{opt}(a)$. For $k \in K_{sim}(a)$, fixed values for $u^{a,k}$ must be given and we add the constraints

$$\left(u^{a,k}\right)^2 \Delta h^{a,k} = \phi^{a,k}, \quad \forall a \in A_{gv} \ \forall k \in K_{sim}(a) \tag{5.14}$$

to our model.

In the other case, i.e., for $k \in K_{opt}(a)$, we add the variables $d_{pos}^{a,k} \in \{0,1\}$ and $d_{neg}^{a,k} \in \{0,1\}$ for $a \in A_{gv}$ and for $k \in K_{opt}(a)$ to our model. A value of $d_{pos}^{a,k} = 1$ indicates non-negative flow and $d_{neg}^{a,k} = 1$ indicates non-positive flow and vice versa. Finally we add the following linear constraints to our model:

$$d_{pos}^{a,k} + d_{neg}^{a,k} \leq 1, \quad \forall a \in A_{gv} \ \forall k \in K_{opt}(a), \tag{5.15a}$$

$$[q_L^{a,k}]^- d_{neg}^{a,k} \leq q_L^{a,k}, \quad \forall a \in A_{gv} \ \forall k \in K_{opt}(a), \tag{5.15b}$$

$$[q_L^{a,k}]^+ d_{pos}^{a,k} \geq q_L^{a,k}, \quad \forall a \in A_{gv} \ \forall k \in K_{opt}(a), \tag{5.15c}$$

$$[\Delta h^{a,k}]^- \left(1 - d_{pos}^{a,k}\right) \leq \Delta h^{a,k}, \quad \forall a \in A_{gv} \ \forall k \in K_{opt}(a), \tag{5.15d}$$

$$[\Delta h^{a,k}]^+ \left(1 - d_{neg}^{a,k}\right) \geq \Delta h^{a,k}, \quad \forall a \in A_{gv} \ \forall k \in K_{opt}(a), \tag{5.15e}$$

$$\phi^{a,k} \leq [\Delta h^{a,k}]^+ d_{pos}^{a,k}, \quad \forall a \in A_{gv} \ \forall k \in K_{opt}(a), \quad (5.15\text{f})$$

$$\phi^{a,k} \geq [\Delta h^{a,k}]^- d_{neg}^{a,k}, \quad \forall a \in A_{gv} \ \forall k \in K_{opt}(a), \quad (5.15\text{g})$$

$$\Delta h^{a,k} - \phi^{a,k} \geq [\Delta h^{a,k}]^- \left(1 - d_{pos}^{a,k}\right), \quad \forall a \in A_{gv} \ \forall k \in K_{opt}(a), \quad (5.15\text{h})$$

$$\Delta h^{a,k} - \phi^{a,k} \leq [\Delta h^{a,k}]^+ \left(1 - d_{neg}^{a,k}\right), \quad \forall a \in A_{gv} \ \forall k \in K_{opt}(a), \quad (5.15\text{i})$$

$$d_{pos}^{a,k}, d_{neg}^{a,k} \in \{0, 1\}, \quad \forall a \in A_{gv} \ \forall k \in K_{opt}(a). \quad (5.15\text{j})$$

Constraints (5.15a)–(5.15c) define the variables $d_{pos}^{a,k}$ and $d_{neg}^{a,k}$ as stated above. Constraints (5.15d)–(5.15e) assert that the sign of the change in the pressure head is in accordance with the flow direction. For a closed valve, inlet and outlet pressure are decoupled. Finally, Constraints (5.15f)–(5.15i) establish the set of feasible points of our gate valve model (5.15) is equal to the solution set of Equation (5.12) with $u^{a,k} \in [0, 1]$, projected on the $(\Delta h^{a,k}, \phi^{a,k})$-plane. Note that the opening degree $u^{a,k}$ can be easily computed a-posteriorily from a solution to (5.15).

Check valves. For a check valve $a \in A_{xv} \subseteq A_{va} \subseteq A$, we are given its friction coefficient ζ_a. A check valve closes whenever the outlet pressure exceeds the inlet pressure to prevent backflow. In any other case it is (fully) open. The change in pressure head that water is affected by passing the valve is again given by

$$u^{a,k} \Delta h^{a,k} = \phi^{a,k}, \quad (5.16)$$

with

$$\phi^{a,k} = \zeta_a \left(q_L^{a,k}\right)^2, \quad \forall a \in A_{xv} \ \forall k \in K, \quad (5.17\text{a})$$

$$\phi^{a,k} \in \mathbb{R}, \quad \forall a \in A_{xv} \ \forall k \in K, \quad (5.17\text{b})$$

and $u^{a,k} \in \{0, 1\}$, which is introduced to our model as an additional variable for all $a \in A_{xv}$ and $k \in K$.

Our check valve model then consists of the constraints

$$q_L^{a,k} \geq 0, \quad \forall a \in A_{xv} \ \forall k \in K, \quad (5.17\text{c})$$

$$[q_L^{a,k}]^+ u^{a,k} \geq q_L^{a,k}, \quad \forall a \in A_{xv} \ \forall k \in K, \quad (5.17\text{d})$$

$$[\Delta h^{a,k}]^- \left(1 - u^{a,k}\right) \leq \Delta h^{a,k}, \quad \forall a \in A_{xv} \ \forall k \in K, \quad (5.17\text{e})$$

$$\Delta h^{a,k} - \phi^{a,k} \geq [\Delta h^{a,k}]^- \left(1 - u^{a,k}\right), \quad \forall a \in A_{xv} \ \forall k \in K, \quad (5.17\text{f})$$

$$\Delta h^{a,k} - \phi^{a,k} \leq [\Delta h^{a,k}]^+ \left(1 - u^{a,k}\right), \quad \forall a \in A_{xv} \ \forall k \in K, \quad (5.17\text{g})$$

$$u^{a,k} \in \{0, 1\}, \quad \forall a \in A_{xv} \ \forall k \in K. \quad (5.17\text{h})$$

Control valves. For a control valve $a = (i, j) \in A_{cv} \subseteq A_{va} \subseteq A$, we are given the area of its inlet A_a, when it is fully open and its friction coefficient ζ_a as in case of a gate valve. To model a control valve, we have to guarantee that the equations

$$\left(u^{a,k}\right)^2 \Delta h^{a,k} = \phi^{a,k}, \quad \forall a \in A_{cv} \ \forall k \in K, \tag{5.18}$$

with

$$\phi^{a,k} = \frac{\zeta_a}{2gA_a^2} \left(q_L^{a,k}\right)^2, \quad \forall a \in A_{cv} \ \forall k \in K, \tag{5.19a}$$

$$\phi^{a,k} \in \mathbb{R}, \quad \forall a \in A_{cv} \ \forall k \in K, \tag{5.19b}$$

are satisfied. Here, $u^{a,k} \in [0, 1]$ denotes the opening degree of a at time k.

Again, as in case of gate valves, let $K_{opt}(a) \subseteq K$ be such that the control of a is subject to optimization during the timespans $[k, k+1)$ for $k \in K_{opt}(a)$ and let $K_{sim}(a) := K \setminus K_{opt}(a)$.

To model a control valve during timespans, where its control is subject to optimization, we add

$$q_L^{a,k} \geq 0, \quad \forall a \in A_{cv} \ \forall k \in K_{opt}(a), \tag{5.20a}$$

$$\phi^{a,k} \in [0, \Delta h^{a,k}], \quad \forall a \in A_{cv} \ \forall k \in K_{opt}(a), \tag{5.20b}$$

to our model.

If $K_{sim}(a) \neq \emptyset$, we are given additional data. First, the control variable $\chi^a \in \{h^{a,0}, h^{a,1}, q^{a,1}\}$ of the valve must be specified. Second, a target value $\chi_{set}^{a,k} \geq 0$ for the control variable must be given for all $k \in K_{sim}(a)$. For convenience, we denote the set of control valves with control variable χ by A_{cv}^χ. For $k \in K_{sim}(a)$ and $\chi \in \{h^{a,1}, q^{a,1}\}$ the control valve must follow the control law

$$\frac{\partial u(t)}{\partial t} = \frac{\chi_{set} - \chi(t)}{\alpha} \left(f_{lo}(t)(1 - u(t)) + f_{up}(t) u(t)\right); \ u(0) = u_0, \tag{5.21}$$

with

$$f_{lo}(t) = \frac{\mathrm{sgn}(\chi_{set} - \chi(t)) + 1}{2}, \quad f_{up}(t) = \frac{1 - \mathrm{sgn}(\chi_{set} - \chi(t))}{2}, \tag{5.22}$$

where $\alpha > 0$ defines the range of the control valve. However, in our stationary model $\partial u(t)/\partial t = 0$ and (5.21) simplifies to

$$(\chi_{set} - \chi(t)) \left(f_{lo}(t)(1 - u(t)) + f_{up}(t) u(t)\right) = 0. \tag{5.23}$$

If the target value can be reached, Equation (5.23) is satisfied due to $\chi(t) = \chi_{set}$. If $\chi(t) < \chi_{set}$, we have $f_{lo}(t) = 1$ and $f_{up}(t) = 0$. Thus Equation (5.23) can only be satisfied with $u(t) = 1$. Finally, if we have $\chi(t) > \chi_{set}$, this results in $f_{lo}(t) = 0$ and

$f_{up}(t) = 1$ such that the only solution to Equation (5.23) is given by $u(t) = 0$ in this case.

For $\chi = h^{a,0}$, we have to swap the roles of f_{lo} and f_{up} such that Equation (5.23) becomes

$$(\chi_{set} - \chi(t))\left(f_{up}(t)(1 - u(t)) + f_{lo}(t)\,u(t)\right) = 0. \quad (5.24)$$

In this case we have either $\chi(t) = \chi_{set}$, or $u(t) = 0$ in case of $\chi(t) < \chi_{set}$, or $u(t) = 1$ in case of $\chi(t) > \chi_{set}$.

In addition to Constraints (5.20), we therefore have to incorporate

$$q_L^{a,k} + [q_L^{a,k}]^+ f_{up}^{a,k} \leq [q_L^{a,k}]^+,$$
$$\forall a \in A_{cv}^{h^{a,1}} \cup A_{cv}^{q^{a,1}}, \; \forall k \in K_{sim}(a), \quad (5.25a)$$

$$\phi^{a,k} + [\phi^{a,k}]^+ f_{up}^{a,k} \leq [\phi^{a,k}]^+,$$
$$\forall a \in A_{cv}^{h^{a,1}} \cup A_{cv}^{q^{a,1}}, \; \forall k \in K_{sim}(a), \quad (5.25b)$$

$$\phi^{a,k} - \Delta h^{a,k} \geq ([\phi^{a,k}]^- - [\Delta h^{a,k}]^+)(1 - f_{lo}^{a,k}),$$
$$\forall a \in A_{cv}^{h^{a,1}} \cup A_{cv}^{q^{a,1}}, \; \forall k \in K_{sim}(a), \quad (5.25c)$$

$$q_L^{a,k} + [q_L^{a,k}]^+ f_{lo}^{a,k} \leq [q_L^{a,k}]^+,$$
$$\forall a \in A_{cv}^{h^{a,0}}, \; \forall k \in K_{sim}(a), \quad (5.25d)$$

$$\phi^{a,k} + [\phi^{a,k}]^+ f_{lo}^{a,k} \leq [\phi^{a,k}]^+,$$
$$\forall a \in A_{cv}^{h^{a,0}}, \; \forall k \in K_{sim}(a), \quad (5.25e)$$

$$\phi^{a,k} - \Delta h^{a,k} \geq ([\phi^{a,k}]^- - [\Delta h^{a,k}]^+)(1 - f_{up}^{a,k}),$$
$$\forall a \in A_{cv}^{h^{a,0}}, \; \forall k \in K_{sim}(a), \quad (5.25f)$$

$$\chi^{a,k} + f_{lo}^{a,k}\left(\chi_{set}^{a,k} - [\chi^{a,k}]^-\right) \geq \chi_{set}^{a,k}, \quad \forall a \in A_{cv} \; \forall k \in K_{sim}(a), \quad (5.25g)$$

$$\chi^{a,k} + f_{up}^{a,k}\left(\chi_{set}^{a,k} - [\chi^{a,k}]^+\right) \leq \chi_{set}^{a,k}, \quad \forall a \in A_{cv} \; \forall k \in K_{sim}(a), \quad (5.25h)$$

$$f_{lo}^{a,k} + f_{up}^{a,k} \leq 1, \quad \forall a \in A_{cv} \; \forall k \in K_{sim}(a), \quad (5.25i)$$

$$f_{lo}^{a,k}, f_{up}^{a,k} \in \{0, 1\}, \quad \forall a \in A_{cv} \; \forall k \in K_{sim}(a). \quad (5.25j)$$

into our model.

In cases, where a non-closed control valve has to be operated within a minimum and a maximum opening degree $0 < u_{min}^a$ or $u_{max}^a < 1$ or if bounds on the control target value of the control valve are given, we have to use both, Constraints (5.20) and (5.25), for all $k \in K$. Moreover, we have to add the additional constraints

$$u_{min}^a(1 - f_{lo}^{a,k} - f_{up}^{a,k}) \leq \phi^{a,k} \leq u_{max}^a \Delta h^{a,k}, \quad \forall a \in A_{cv} \; \forall k \in K. \quad (5.26)$$

To take care of a lower bound $[\chi_{set}^{a,k}]^-$ and an upper bound $[\chi_{set}^{a,k}]^+$ on the control target, we replace Constraints (5.25g)–(5.25h) by

$$\chi^{a,k} + f_{lo}^{a,k}([\chi_{set}^{a,k}]^- - [\chi^{a,k}]^-) \\ + f_{up}^{a,k}([\chi_{set}^{a,k}]^- - [\chi_{set}^{a,k}]^+) \geq [\chi_{set}^{a,k}]^-, \quad (5.27)$$

$$\chi^{a,k} + f_{up}^{a,k}([\chi_{set}^{a,k}]^+ - [\chi^{a,k}]^+) \\ + f_{lo}^{a,k}([\chi_{set}^{a,k}]^+ - [\chi_{set}^{a,k}]^-) \leq [\chi_{set}^{a,k}]^+, \quad (5.28)$$

forall $a \in A_{cv}$ and forall $k \in K$.

Pumps. With each pump $a \in A_{pu} \subseteq A$ we associate a relative speed $\omega^a(k)$, where $\omega^a = 1$ is the relative nominal speed. Moreover we are given a relative minimal speed $\omega_{min}^a \leq 1$ and maximal speed $\omega_{max}^a \geq 1$. Note that relative speed values are determined by dividing absolute speed values by the absolute nominal speed.

For each pump a flow dependent head increase curve $\psi^a(q^a)$ corresponding to the nominal speed is defined. Each such curve is assumed to have a generic form $\psi^a(x) = \alpha_0^a + \alpha_r^a x^{r^a}$. The parameters for the nominal speed curve $\alpha_0^a, \alpha_r^a, r^a$ are either given directly or computed by linear interpolation from parameters of the nearest curves for $\omega^a > 1$ and $\omega^a < 1$. To include varying speeds we use the parametric characteristic curve model

$$\Delta h^a(q^a(k), \omega^a(k)) = (\omega^a)^{r^a} \left(\alpha_0^a + \alpha_r^a \left(\frac{q^a(k)}{\omega^a(k)} \right)^{r^a} \right). \quad (5.29)$$

This formula is similar to the one proposed in [3], but differs slightly for $r^a \neq 2$. In particular we use the first affinity law and for $r^a \neq 2$ an approximation of the second affinity law. This approximation has the advantage that no additional nonlinear terms for the relative speed are necessary. For realistic pump curves $\alpha_0^a > 0$, $\alpha_r^a < 0$, and $r^a > 1$, which is assumed for our model.

For all pumps we introduce the additional variables

$$y^{a,k} \in \{0, 1\}, \quad \forall a \in A_{pu} \ \forall k \in K, \quad (5.30a)$$
$$\psi^{a,k} \quad \forall a \in A_{pu} \ \forall k \in K, \quad (5.30b)$$
$$\omega^{a,k} \quad \forall a \in A_{pu} \ \forall k \in K, \quad (5.30c)$$
$$\tilde{\omega}^{a,k} \quad \forall a \in A_{pu} \ \forall k \in K. \quad (5.30d)$$

The binary variables y represent the state of pump, i.e., running (1) or shut off (0) and ψ is an auxiliary variable for the nonlinear part of the pump curve. The variable ω is the relative speed of a pump and $\tilde{\omega}$ is an auxiliary variable representing the relative speed of a pump for active pumps and a fictive relative speed for inactive pumps.

We add the constraints

$$q_L^{a,k} = q_R^{a,k}, \tag{5.31a}$$

$$q_L^{a,k} \geq q_{\min}^a y^{a,k}, \tag{5.31b}$$

$$q_L^{a,k} \leq [q_L^{a,k}]^+ y^{a,k}, \tag{5.31c}$$

$$\psi^{a,k} = \alpha_r^a \left(q_L^{a,k}\right)^{r^a}, \tag{5.31d}$$

$$h_R^{a,k} - h_L^{a,k} = \alpha_0^a \left(\tilde{\omega}^{a,k}\right)^{r^a} + \psi^{a,k}, \tag{5.31e}$$

$$h_R^{a,k} - h_L^{a,k} \geq \left([h_R^{a,k}]^- - [h_L^{a,k}]^+\right)(1 - y^{a,k}), \tag{5.31f}$$

$$\left(\tilde{\omega}_{\min}^a\right)^{r^a}(1 - y^{a,k}) \leq \left(\tilde{\omega}^{a,k}\right)^{r^a} - \left(\omega^{a,k}\right)^{r^a}, \tag{5.31g}$$

$$\left(\tilde{\omega}_{\max}^a\right)^{r^a}(1 - y^{a,k}) \geq \left(\tilde{\omega}^{a,k}\right)^{r^a} - \left(\omega^{a,k}\right)^{r^a}, \tag{5.31h}$$

$$\left(\omega_{\min}^a\right)^{r^a} y^{a,k} \leq \left(\omega^{a,k}\right)^{r^a}, \tag{5.31i}$$

$$\left(\omega_{\max}^a\right)^{r^a} y^{a,k} \geq \left(\omega^{a,k}\right)^{r^a}, \tag{5.31j}$$

$$y^{a,k} \in \{0, 1\}, \tag{5.31k}$$

for each $a \in A_{pu}$ and each $k \in K$ to our model.

Equations (5.31a) simply state the flow conservation along a pump. Inequalities (5.31b) are used to ensure a minimum flow bound for active pumps $q_{\min}^a \geq 0$ and avoids backflow for inactive pumps. Similar, Inequalities (5.31c) are used to block flow in case on inactive pumps. Equations (5.31d) define the nonlinear part of the curves. Equations (5.31e) give the relation between pressure increase and the pump curves for the nominal speed. The auxiliary variables $\tilde{\omega}$ are introduced to enable deviations from a pump curve corresponding to an inactive pump. If a pump is active at a time step k, indicated by the binary variable $y^{a,k} = 1$, then $\omega_{\min}^a \leq \tilde{\omega}^{a,k} = \omega^{a,k} \leq \omega_{\max}$ is enforced by Constraints (5.31f)–(5.31j). Otherwise for $y^{a,k} = 0$ these constraints are redundant; at least for suitably chosen constants $\tilde{\omega}_{\min}$ and $\tilde{\omega}_{\max}$. In the case of an active pump at time step k, $y^{a,k} = 1$, due to $\tilde{\omega}^{a,k} = \omega^{a,k}$ pressure increase is forced to satisfy the parametric curve (5.29). Finally, Inequalities (5.31f) are added to avoid any negative pressure increase in case of active pumps and are redundant otherwise.

We remark that this model has the advantage that only a single nonlinear term, in particular the right-hand-side of Equation (5.31d), for each pump and timestep is required. Since the variables $\omega^{a,k}$ and $\omega^{a,k}$ arise only in the terms $\left(\omega^{a,k}\right)^{r^a}$ and $\left(\tilde{\omega}^{a,k}\right)^{r^a}$ a respective adapted variable is sufficient.

To respect minimum run time r^a_{min} and minimum down time d^a_{min} restrictions of pumps the additional constraints

$$y^{a,k} - y^{a,k-1} \leq y^{a,l} \quad \forall l \in \left\{k+1,\ldots,k+\left\lceil\frac{r^a_{min}}{\Delta t}\right\rceil - 1\right\} \cap K, \quad (5.32a)$$

$$y^{a,k-1} - y^{a,k} \leq 1 - y^{a,l} \quad \forall l \in \left\{k+1,\ldots,k+\left\lceil\frac{d^a_{min}}{\Delta t}\right\rceil - 1\right\} \cap K, \quad (5.32b)$$

for each $a \in A_{pu}$ and $k \in K$ are included. We use the ceiling of $r^a_{min}/\Delta t$ and $d^a_{min}/\Delta t$ to guarantee minimum run/down times in case of a fractional number of time steps.

Pumps with variable speed. For variable speed pumps it may be necessary to respect a further given control law similar to the one used for control valves. The control law is given by

$$\frac{\partial \omega(t)}{\partial t} = \frac{\chi_{set} - \chi(t)}{\alpha}\left(f_{lo}(t)(\omega_{max} - \omega(t)) + f_{up}(t)(\omega(t) - \omega^a_{min}))\right), \quad (5.33)$$

with

$$f_{lo}(t) = \frac{\text{sgn}(\chi_{set} - \chi(t)) + 1}{2}, \quad f_{up}(t) = \frac{1 - \text{sgn}(\chi_{set} - \chi(t))}{2}, \quad (5.34)$$

and initial condition $\omega(0) = \omega_0$.

In our stationary model $\partial \omega(t)/\partial t = 0$ and this control law simplifies to

$$\begin{cases} \omega(t) \in [\omega_{min}, \omega_{max}], & \text{if } \chi(t) = \chi_{set}, \\ \omega(t) = \omega_{max}, & \text{if } \chi(t) < \chi_{set}, \\ \omega(t) = \omega_{min}, & \text{if } \chi(t) > \chi_{set}. \end{cases} \quad (5.35)$$

In our case we assume a control law with $\chi^a = h^{a,1}$ which is trying to reach a given target outlet pressure $\chi^{a,k}_{set}$.

In the following we denote the set of variable speed pumps with target outlet pressure by $A^{\chi_{set}}_{pv}$. As in case of gate and controle valves, let $K_{opt}(a) \subseteq K$ be such that the control of a is subject to optimization during the timespans $[k, k+1)$ for $k \in K_{opt}(a)$ and let $K_{sim}(a) := K \setminus K_{opt}(a)$ denote the complement.

For the control law we add the variables

$$f^{a,k}_{lo} \in \{0,1\}, \quad a \in A^{\chi_{set}}_{pv} \; k \in K_{sim}(a), \quad (5.36a)$$

$$f^{a,k}_{up} \in \{0,1\}, \quad a \in A^{\chi_{set}}_{pv} \; k \in K_{sim}(a). \quad (5.36b)$$

Here $f_{lo} = 1$ implies that the outlet pressure is not above the target value and the pump it at its maximum speed. Analogously, $f_{up} = 1$ implies that the outlet pressure is not below the target value and the pump it at its minimum speed.

To reflect the control law the constraints

$$h_R^{a,k} + \left(\chi_{set}^{a,k} - [h_R^{a,k}]^-\right)(f_{up}^{a,k} - y^{a,k}) \geq [h_R^{a,k}]^-, \tag{5.37a}$$

$$h_R^{a,k} + \left(\chi_{set}^{a,k} - [h_R^{a,k}]^+\right)(f_{lo}^{a,k} - y^{a,k}) \leq [h_R^{a,k}]^+, \tag{5.37b}$$

$$\left(\tilde{\omega}^{a,k}\right)^{r^a} + \left(\left([\tilde{\omega}^{a,k}]^+\right)^{r^a} - (\omega_{min}^a)^{r^a}\right)f_{lo}^{a,k} \leq \left([\tilde{\omega}^{a,k}]^+\right)^{r^a}, \tag{5.37c}$$

$$\left(\tilde{\omega}^{a,k}\right)^{r^a} + \left(\left([\tilde{\omega}^{a,k}]^-\right)^{r^a} - (\omega_{max}^a)^{r^a}\right)f_{up}^{a,k} \geq \left([\tilde{\omega}^{a,k}]^-\right)^{r^a}, \tag{5.37d}$$

$$f_{lo}^{a,k} + f_{up}^{a,k} \leq y^{a,k}, \tag{5.37e}$$

$$f_{lo}^{a,k}, f_{up}^{a,k} \in \{0,1\}, \tag{5.37f}$$

for each $a \in A_{pv}^{\chi_{set}} \subseteq A_{pv}$ and $k \in K_{sim}(a)$ are added.

For the binary assignment $y^{a,k} = 1$, $f_{lo}^{a,k} = f_{up}^{a,k} = 0$ the Constraints (5.37a) and (5.37b) ensure that $h_R^{a,k} = \chi_{set}^{a,k}$, while the Constraints (5.37c) and (5.37d) become redundant. For $y^{a,k} = f_{lo}^{a,k} = 1$, $f_{up}^{a,k} = 0$ the Constraints (5.37a) and (5.37b) guarantee that $h_R^{a,k} \leq \chi_{set}^{a,k}$ and the Constraints (5.37c) and (5.37d) force $\tilde{\omega}^{a,k} = \omega_{min}^a$. Similar for $y^{a,k} = f_{up}^{a,k} = 1$, $f_{lo}^{a,k} = 0$ we obtain $h_R^{a,k} \geq \chi_{set}^{a,k}$ and $\tilde{\omega}^{a,k} = \omega_{max}^a$. For an inactive pump all constraints (5.37a)–(5.37d) are redundant. Further assignments for $y^{a,k}$, $f_{lo}^{a,k}$, $f_{up}^{a,k}$ are excluded by the Constraints (5.37e)–(5.37f).

Pumps with fixed speed. The same model as for fixed speed pumps is used. For fixed speed pumps $a \in A_{pf} \subseteq A_{pu} \subseteq A$ we simply choose $\omega_{min}^a = \omega_{max}^a = 1$.

Connections. For each connection $a \in A_{co} \subseteq A$ we are given a generic flow dependent curve $\psi^a(q^a)$. The generic curve $\psi^a(x) = \alpha_0^a + \alpha_1^a x + \alpha_r^a x^{r^a}$ is defined by connection specific parameters $\alpha_0^a, \alpha_1^a, \alpha_r^a, r^a$.

A connection can have different types. It can either be a hydraulic coupled element with an associated generic flow dependent pressure head loss curve denoted by $a \in A_{ch} \subset A_{co} \subseteq A$ or a hydraulic decoupled element with an associated generic flow dependent inlet pressure head curve denoted by $a \in A_{cn} \subset A_{co} \subseteq A$.

Moreover all connections are flow conservative.

Thus for all connections the equations

$$q_L^{a,k} = q_R^{a,k}, \quad \forall a \in A_{co} \; \forall k \in K \tag{5.38a}$$

$$\psi^{a,k} = \alpha_0^a + \alpha_1^a q_L^{a,k} + \alpha_r^a \left(q_L^{a,k}\right)^{r^a} \quad \forall a \in A_{co} \; \forall k \in K \tag{5.38b}$$

are added.

Connections with hydraulic coupled pressure loss. A hydraulic coupled connection with an associated generic flow dependent pressure head loss curve $a \in A_{ch} \subset A_{co} \subseteq A$ has the additional constraints

$$h_L^{a,k} - h_R^{a,k} = \psi^{a,k}, \quad \forall a \in A_{ch} \; \forall k \in K. \tag{5.39}$$

Connections with hydraulic decoupled inlet pressure. A hydraulic coupled connection with an associated generic flow dependent pressure head loss curve $a \in A_{ch} \subset A_{co} \subseteq A$ has the additional constraints

$$h_L^{a,k} = \psi^{a,k}, \quad \forall a \in A_{cn} \; \forall k \in K, \tag{5.40a}$$

$$h_L^{a,k} \geq h_R^{a,k}, \quad \forall a \in A_{cn} \; \forall k \in K. \tag{5.40b}$$

Energy aspects and objective function

Our goal is to minimize the amount of total energy consumption, more precisely the total energy costs, required to operate a water supply network for a given scenario (defined by demands, presets, time horizon etc.).

In general pumps are the main energy consumers within a water supply network, but there may be further energy consumption by components like i.e., UV radiation, which could be reflected by a connection in our model.

To handle generic energy consumption we associate with each pump and connection a flow-dependent energy consumption $P(t) \in \mathbb{R}$, with

$$P(q(t), t) = \iota(t) f(q(t)), \tag{5.41}$$

where

$$f(q(t), t) = \begin{cases} \beta_0 + \beta_1 q(t) + \beta_e q(t)^e & \text{or} \\ \text{piecewise constant} \end{cases}. \tag{5.42}$$

The indicator function $\iota : t \to \{0, 1\}$ is to reflect whether the element is active and there is energy consumption or not. Additional binary variables can be introduced, which must be then inevitably coupled to existing variables. In general the implication $\iota(t) = 0 \implies q(t) = 0$ should be satisfied to reflect realistic behavoir or $\iota(t)$ must be coupled to existing control variables.

For ease of presentation we assume that for pumps $\iota(t)$ is defined by the binary decision $y(t)$ indicating whether or not the pump is running and $\iota(t) = 1$ for connections. Moreover we assume that only parametric curves and no piecewise constant function are allowed for pumps.

Altogether we add power consumption variables

$$P^{a,k} \in \mathbb{R}, \quad \forall a \in A_{pu} \cup A_{co} \; \forall k \in K, \tag{5.43}$$

and the constraints

$$P^{a,K} = \beta_0^a y^{a,k} + \beta_1^a q^{a,k} + \beta_e^a q^{a,k} \quad \forall a \in A_{pu} \ \forall k \in K, \quad (5.44a)$$

$$P^{a,K} = \begin{cases} \beta_0^a + \beta_1^a q^{a,k} + \beta_e^a q^{a,k} & \text{or} \\ \text{piecewise constant} \end{cases} \quad \forall a \in A_{co} \ \forall k \in K, \quad (5.44b)$$

to our model.

Total energy costs can then be written as

$$\sum_{a \in A_{pu} \cup A_{co}} \sum_{k \in K} c^{a,k} P^{a,k} \Delta t \quad (5.45)$$

We remark that negative power consumption can be used to incorporate energy recovery.

Model summary. Our MINLP model can be summarized as follow

$$\begin{aligned}
& \min \quad (5.45) & (5.46a) \\
& \text{s.t.} \quad \text{junctions: } (5.4a) - (5.4c) & (5.46b) \\
& \quad \text{tanks: } (5.5a) - (5.5f) & (5.46c) \\
& \quad \text{pipes: } (5.9a) - (5.10) & (5.46d) \\
& \quad \text{valves: } (5.11), (5.13), (5.14), (5.15), (5.17), (5.19), (5.20), (5.25) & (5.46e) \\
& \quad \text{pumps: } (5.31), (5.32), (5.37) & (5.46f) \\
& \quad \text{connections: } (5.38), (5.39), (5.40) & (5.46g) \\
& \quad \text{power consumption: } (5.44) & (5.46h)
\end{aligned}$$

5.2.2 Solution approach

We can not expect to solve the stationary MINLP model (5.46) for networks of realistic size within reasonable computing time with the aid of even the best available general-purpose MINLP solver at this point in time. We therefore conclude that a tailored algorithm has to be developed in order to fulfill all performance and accuracy requirements that are needed for real-world daily operation.

The approach presented in the following is based on the idea of decomposing a water network into passive and active subnetworks and to then exploit this decompostion by a penalty alternating direction method (PADM) to solve the stationary MINLP (5.46).

Network decomposition. In order to describe our decomposition, we start with some definitions.

Definition 1. Let $G = (V, E)$ be an undirected graph. Then the *connected components* of G are the subgraphs $G' = (V', E')$ of G with maximal sets of vertices $V' \subseteq V$ that are pairwisely connected by a path and $E' = \{\{u, v\} \in E : u, v \in V'\}$.

Definition 2. Let $G = (V, A)$ be a directed graph not containing antiparallel or parallel edges and let $G' = (V, E)$ be the undirected graph, which contains an (undirected) edge $\{u, v\} \in E$ if and only if $(u, v) \in A$ or $(v, u) \in A$. Further, let

$$C'_1 = (V_1, E_1), \ldots, C'_c = (V_c, E_c)$$

be the connected components of G'. Then we call

$$C_1 = (V_1, A_1), \ldots, C_c = (V_c, A_c)$$

the *connected components* of G, where $a = (i, j) \in A_k$ if and only if $\{i, j\} \in E_k$ for all $a \in A$ and $k = 1, \ldots, c$.

Definition 3. Given the graph $G = (V, A)$ of a water network, let $A_{act} = A_{va} \cup A_{pu} \subseteq A$ be the *active edges* of G and let $A_{pas} = A \setminus A_{act}$ be the *passive edges* of G. Then the connected components of the graph $G' = (V, A \setminus A_{act})$ are the *passive subnetworks* of G and $G'' = (V, A \setminus A_{pas})$ are the *active subnetworks* of G.

Definition 4. Let $G = (V, A)$ be the graph of a water network. Let

$$\mathcal{A} = \{G_{act,1} = (V_{act,1}, A_{act,1}), \ldots, G_{act,c_{act}} = (V_{act,1}, A_{act,c_{act}})\}$$

be the active components and let

$$\mathcal{P} = \{G_{pas,1} = (V_{pas,1}, A_{pas,1}), \ldots, G_{pas,c_{pas}} = (V_{pas,1}, A_{pas,c_{pas}})\}$$

be the passive components of G. Then we call any node

$$i \in \left(\bigcup_{k=1}^{c_{act}} V_{act,k}\right) \cap \left(\bigcup_{k=1}^{c_{pas}} V_{pas,k}\right)$$

a *coupling node*.

As one can already see from the above definitions, the active (passive) subnetworks of a water network can be obtained by removing the passive (active) edges from the graph and by then computing the connected components of the remaining graph. The latter can be done in linear time by, e.g. breadth-first search.

Observe that passive subnetworks mainly consist of pipes and that each controllable element is part of an active subnetwork due to the decomposition. Further note that a water network typically consists of a small number of large passive (pipe) subnetworks and some active subnetworks, whereof the largest ones are typically constituted by the waterworks.

This means that once we are able to solve the entire model by computing solutions for the subnetworks one by one, the problems that remain to be solved would be MINLPs on the scale of waterworks and NLPs on the scale of the pipe network, instead of one large MINLP on the scale of the entire network. This would mean a tremendous reduction in computational complexity.

In the remainder of this chapter we show how a decomposition into active and passive subnetworks can be exploited in the desired way by an algorithm that is called the Penalty Alternating Direction Method.

Algorithm 1: A standard alternating direction method
---
1 Choose initial values $(x^0, y^0) \in X \times Y$.
2 **for** $k = 0, 1, \ldots$ **do**
3 $\quad$ Compute
      $x^{k+1} \in \mathrm{argmin}_x\{f(x, y^k) : g(x, y^k) = 0, \ h(x, y^k) \geq 0, \ x \in X\}$.
4 $\quad$ Compute $\ y^{k+1} \in \mathrm{argmin}_y\{f(x^{k+1}, y) : g(x^{k+1}, y) = 0, \ h(x^{k+1}, y) \geq 0, \ y \in Y\}$.
5 $\quad$ Set $k \leftarrow k + 1$

Alternating directon methods (ADMs). Before we actually introduce the penalty alternating direction method (PADM), we briefly review classical alternating direction methods (ADMs), cf. [10]. To this end, we consider the general problem

$$\min_{x,y} \ f(x, y) \tag{5.47a}$$

$$\text{s.t.} \quad g(x, y) = 0, \quad h(x, y) \geq 0, \tag{5.47b}$$

$$x \in X, \quad y \in Y, \tag{5.47c}$$

for which we make the following assumption:

Assumption 1. *The objective function* $f : \mathbb{R}^{n_x+n_y} \to \mathbb{R}$ *and the constraint functions* $g : \mathbb{R}^{n_x+n_y} \to \mathbb{R}^m$, $h : \mathbb{R}^{n_x+n_y} \to \mathbb{R}^p$ *are continuous and the sets X and Y are non-empty and compact.*

The feasible set is denoted by Ω, i.e.,

$$\Omega = \{(x, y) \in X \times Y : g(x, y) = 0, \ h(x, y) \geq 0\} \subseteq X \times Y,$$

and the corresponding projections onto X and Y are denoted by Ω_X and Ω_Y, respectively. Classical alternating direction methods are extensions of Lagrangian methods and have been originally proposed in Gabay and Mercier [6] and Glowinski and Marroco [12]. More recently, ADM-type methods have seen a resurgence; see, e.g., [2] for a general overview and [11, 9] for applications of ADMs to nonconvex MINLPs from gas transport.

ADMs solve Problem (5.47) by solving two simpler problems: Given an iterate (x^k, y^k) they solve Problem (5.47) for y fixed to y^k into the direction of x, yielding a new x-iterate x^{k+1}. Afterward, x is fixed to x^{k+1} and Problem (5.47) is solved into the direction of y, yielding a new y-iterate y^{k+1}. A formal listing is given in Algorithm 1.

If the optimization problem in Line 3 or Line 4 of Algorithm 1 has a unique solution for all k, it is known that ADMs converge to so-called *partial minima* of Problem (5.47), i.e., to points $(x^*, y^*) \in \Omega$ for which

$$f(x^*, y^*) \leq f(x, y^*) \quad \text{for all } (x, y^*) \in \Omega,$$
$$f(x^*, y^*) \leq f(x^*, y) \quad \text{for all } (x^*, y) \in \Omega$$

holds; see Gorski, Pfeuffer, and Klamroth [13] for the following result:

Theorem 1. *Let $\{(x^i, y^i)\}_{i=0}^{\infty}$ be a sequence with $(x^{i+1}, y^{i+1}) \in \Theta(x^i, y^i)$, where*

$$\Theta(x^i, y^i) := \{(x^*, y^*) : \forall x \in X. \ f(x^*, y^i) \\ \leq f(x, y^i); \ \forall y \in Y. \ f(x^*, y^*) \leq f(x^*, y)\}.$$

Suppose that Assumption 1 holds and that the solution of the first optimization problem is always unique. Then every convergent subsequence of $\{(x^i, y^i)\}_{i=0}^{\infty}$ converges to a partial minimum. For two limit points z, z' of such subsequences it holds that $f(z) = f(z')$.

Stronger results can be obtained if additional assumptions are made on f and Ω: If f is continuously differentiable, Algorithm 1 converges to a stationary point (in the classical sense of nonlinear optimization). If, in addition, f and Ω are convex it is easy to show that partial minimizers are also global minimizers of Problem (5.47). For more details on the convergence theory of classical ADMs, see Gorski, Pfeufer, and Klamroth [13] as well as Wendell and Hurter [33].

Penalty alternating directon methods (PADMs). We now present the weighted ℓ_1 penalty alternating direction method (PADM) from [10]. To this end, we define the ℓ_1 penalty function

$$\phi_1(x, y; \mu, \rho) := f(x, y) + \sum_{i=1}^{m} \mu_i |g_i(x, y)| + \sum_{i=1}^{p} \rho_i [h_i(x, y)]^-,$$

where $[\alpha]^- := \max\{0, -\alpha\}$ holds and $\mu = (\mu_i)_{i=1}^m, \rho = (\rho_i)_{i=1}^p \geq 0$ are the penalty parameters for the equality and inequality constraints. Note that we allow for different penalty parameters for the constraints instead of a single penalty parameter as it is often the case for penalty methods.

The penalty ADM now proceeds as follows. Given a starting point and initial values for all penalty parameters, the alternating direction method of Algorithm 1 is used to compute a partial minimum of the penalty problem

$$\min_{x,y} \ \phi_1(x, y; \mu, \rho) \quad \text{s.t.} \quad x \in X, \ y \in Y. \tag{5.48}$$

Afterward, the penalty parameters are updated and the next penalty problem is solved to partial minimality. Thus, the algorithm produces a sequence of partial minima of a sequence of penalty problems of type (5.48). More formally, the method is specified in Algorithm 2. Next, we present the most important convergence results for the PADM. However, the detailed proofs are omitted but can be found in [10]. We start by showing that partial minima of the penalty problems are partial minima of the original problem if they are feasible.

Algorithm 2: The ℓ_1 penalty alternating direction method

1 Choose initial values $(x^{0,0}, y^{0,0}) \in X \times Y$ and penalty parameters $\mu^0, \rho^0 \geq 0$.
   for $k = 0, 1, \ldots$ **do**
2   Set $l = 0$
3   **while** $(x^{k,l}, y^{k,l})$ is not a partial minimum of (5.48) with $\mu = \mu^k$ and $\rho = \rho^k$ **do**
4   Compute $x^{k,l+1} \in \mathrm{argmin}_x \{\phi_1(x, y^{k,l}; \mu^k, \rho^k) : x \in X\}$.
5   Compute $y^{k,l+1} \in \mathrm{argmin}_y \{\phi_1(x^{k,l+1}, y; \mu^k, \rho^k) : y \in Y\}$.
6   Set $l \leftarrow l + 1$.
7   Choose new penalty parameters $\mu^{k+1} \geq \mu^k$ and $\rho^{k+1} \geq \rho^k$.

Lemma 1. *Assume that (x^*, y^*) is a partial minimum of $\phi_1(x, y; \mu, \rho)$ for arbitrary but fixed $\mu, \rho \geq 0$ and let (x^*, y^*) be feasible for Problem (5.47). Then (x^*, y^*) is a partial minimum of Problem (5.47).*

For the next theorem we need some more notation. Let χ be the ℓ_1 feasibility measure of Problem (5.47), which we define as

$$\chi(x, y) := \sum_{i=1}^{m} |g_i(x, y)| + \sum_{i=1}^{p} [h_i(x, y)]^-.$$

Obviously, $\chi(x, y) \geq 0$ holds and $\chi(x, y) = 0$ if and only if (x, y) is feasible w.r.t. g and h. Moreover, we define the weighted ℓ_1 feasibility measure as

$$\chi_{\mu,\rho}(x, y) := \sum_{i=1}^{m} \mu_i |g_i(x, y)| + \sum_{i=1}^{p} \rho_i [h_i(x, y)]^-,$$

i.e., our ℓ_1 penalty function can be stated as

$$\phi_1(x, y; \mu, \rho) = f(x, y) + \chi_{\mu,\rho}(x, y).$$

The next theorem states that the sequence of partial minima of the iteratively solved penalty problems converges to a partial minimum of $\chi_{\mu,\rho}$.

Lemma 2. *Suppose that Assumption 1 holds and that $\mu_i^k \nearrow \infty$ for all $i = 1, \ldots, m$ and $\rho_i^k \nearrow \infty$ for all $i = 1, \ldots, p$. Moreover, let (x^k, y^k) be a sequence of partial minima of (5.48) (for $\mu = \mu^k$ and $\rho = \rho^k$) generated by Algorithm 2 with $(x^k, y^k) \to (x^*, y^*)$. Then there exist weights $\bar{\mu}, \bar{\rho} \geq 0$ such that (x^*, y^*) is a partial minimizer of the feasibility measure $\chi_{\bar{\mu},\bar{\rho}}$.*

The two preceding lemmas now enable us to characterize the overall convergence behavior of the penalty ADM algorithm 2.

Theorem 2. *Suppose that Assumption 1 holds and that $\mu_i^k \nearrow \infty$ for all $i = 1, \ldots, m$ and $\rho_i^k \nearrow \infty$ for all $i = 1, \ldots, p$. Moreover, let (x^k, y^k) be a sequence of partial minima of (5.48) (for $\mu = \mu^k$ and $\rho = \rho^k$) generated by Algorithm 2 with $(x^k, y^k) \to (x^*, y^*)$. Then there exist weights $\bar{\mu}, \bar{\rho} \geq 0$ such that (x^*, y^*) is a partial minimizer of the feasibility measure $\chi_{\bar{\mu}, \bar{\rho}}$.*

If, in addition, (x^*, y^*) is feasible for the original problem (5.47), the following holds:

1. *If f is continuous, then (x^*, y^*) is a partial minimum of (5.47).*
2. *If f is continuously differentiable, then (x^*, y^*) is a stationary point of (5.47).*
3. *If f is continuously differentiable and f and Ω are convex, then (x^*, y^*) is a global optimum of (5.47).*

In the next theorem we generalize the classical result on the exactness of the ℓ_1 penalty function (see, e.g., [15, 22]) to the setting of partial minima. For the ease of presentation, we state this result only for the case without inequality constraints. However, the result can also be applied to problems including inequality constraints by using standard reformulation techniques to translate inequality constrained to equality constrained problems. Beforehand, we need two assumptions:

Assumption 2. *The objective function $f : X \times Y \to \mathbb{R}$ of Problem (5.47) is locally Lipschitz continuous in the direction of x and of y, i.e., for every $(x^*, y^*) \in \Omega$ there exists an open set $N(x^*, y^*)$ containing (x^*, y^*) and a constant $L \geq 0$ such that*

$$|f(x, y^*) - f(x^*, y^*)| \leq L\|x - x^*\| \quad \text{for all } x \text{ with } (x, y^*) \in N(x^*, y^*),$$
$$|f(x^*, y) - f(x^*, y^*)| \leq L\|y - y^*\| \quad \text{for all } y \text{ with } (x^*, y) \in N(x^*, y^*).$$

Note that if one set, say Y, is discrete, the corresponding condition is trivially satisfied. In this case any set of the form $(U, \{y^*\})$, where $U \subseteq X$ is an open neighborhood around x^*, is an open neighborhood around (x^*, y^*).

Assumption 3. *For every constraint $g_i, i = 1, \ldots, m$, there exists a constant $l_i > 0$ such that*

$$l_i \|x - x^*\| \leq |g_i(x, y^*) - g_i(x^*, y^*)| \quad \text{for all } x \text{ with } (x, y^*) \in N(x^*, y^*),$$
$$l_i \|y - y^*\| \leq |g_i(x^*, y) - g_i(x^*, y^*)| \quad \text{for all } y \text{ with } (x^*, y) \in N(x^*, y^*).$$

Note that in the case of existing directional derivatives of g_i, the latter assumption states that the directional derivatives of the g_i both in the direction of x and of y are bounded away from zero. Before we state and proof the exactness theorem we briefly discuss the latter assumption. In the context of ADMs, the constraints $g(x, y) = 0$ are mostly so-called copy constraints of the type

$$g(x, y) = A(x - y) = 0$$

that are used to decompose the genuine problem formulation such that it fits into the framework of Problem (5.47); see, e.g., Nowak [23]. If the matrix A is square and has full rank—as it is typically the case for copy constraints—the constraints g are bi-Lipschitz and thus fulfill Assumption 3. Now, we are ready to state the theorem on exactness of the ℓ_1 penalty function w.r.t. partial minima.

Theorem 3. *Let (x^*, y^*) be a partial minimizer of*

$$\min_{x,y} \quad f(x,y) \quad \text{s.t.} \quad g(x,y) = 0, \ x \in X, \ y \in Y, \tag{5.49}$$

and suppose that Assumptions 2 and 3 hold. Then there exists a constant $\bar{\mu} > 0$ such that (x^*, y^*) is a partial minimizer of

$$\min_{x,y} \quad \phi_1(x,y;\mu) \quad \text{s.t.} \quad x \in X, \ y \in Y$$

for all $\mu \geq \bar{\mu}$ and

$$\phi_1(x,y;\mu) := f(x,y) + \sum_{i=1}^{m} \mu_i \, |g_i(x,y)|.$$

PADM for water network optimization. To solve Model (5.46) for a water network $G = (V, A)$ by Algorithm 2, we first compute the active and passive components of G. For the sake of simplicity let us assume that G decomposes into one active component $G^A = (V^A, A^A)$ and one passive component $G^P = (V^P, A^P)$ with coupling nodes $C \subseteq V$. W.l.o.g., we additionally assume $C \subseteq V_{ju}$ and $q_{ext}^{v,k} = \text{const}$ for all $v \in C$.

Next, we replace each node $v \in C \cap V^A$ by a new node v^A connected to an unbounded external source and each edge $a \in A^A$ that starts (ends) at node v is replaced by a new edge that starts (ends) in node v^A. Analogously, we replace each node $v \in C \cap V^P$ by a new node v^P connected to an unbounded external source and each edge $a \in A^P$ that starts (ends) at node v is replaced by a new edge that starts (ends) in node v^P. This essentially means a doubling of each coupling node, where the newly introduced unbounded external source represents the flow into or out of the respective component.

Now we can write Model (5.46) in the form (5.47), where $f(x,y)$ is the objective function (5.45), X is defined by the set of constraints of Model (5.46) for G^A, Y is defined by the set of constraints of Model (5.46) for G^P, and $g(x,y)$ is replaced by

$$\begin{pmatrix} h^{v^A,k} - h^{v^P,k} \\ q_{ext}^{v^A,k} + q_{ext}^{v^P,k} + q_{ext}^{v,k} \end{pmatrix} \quad \forall v \in C \ \forall k \in K \tag{5.50}$$

and there are no coupling inequalities $h(x,y) \geq 0$. In case of variable external flow, one would have to decide, whether $q_{ext}^{v,k}$ is contained in the x- or in the y-variables from Model 5.47.

Finally, we would like to remark that in case of multiple active components, X is the union of sets in disjoint dimensions, which means that the optimization in Line 4 of Algorithm 2 can be performed by solving for the active subnetworks independently. Analogously, the same holds in case of multiple passive subnetworks.

Solving the active subnetwork problem. As it turned out that it might still take hours to solve some of the MINLPs that correspond to waterworks, the overall performance of our PADM algorithm from Section 5.2.2 is not yet satisfactory for real-life networks. To overcome this problem we pursue deriving an a priori mixed-integer linear relaxation by means of piecewise linear approximations to nonlinear functions. To this end let us consider Model (5.46) for an active component of a water network in the following abstract form:

$$\min_{x} \quad c^T x \tag{5.51a}$$

$$\text{s.t.} \quad a_i^T x + \sum_{j=1}^{T_i} f_{i,j}(x) = 0 \quad i = 1, \ldots, m, \tag{5.51b}$$

$$x \in \mathbb{R}^{n-p} \times \mathbb{Z}^p, \tag{5.51c}$$

with nonlinear functions $f_{i,j} : \mathbb{R}^n \to \mathbb{R}$ for $i = 1, \ldots, m$ and $j = 1, \ldots, T_i$. The basic idea from [7] and [8] for such relaxations is as follows (cf. [11]): To obtain a mixed-integer linear relaxation of (5.51), we first replace each occurrence of a nonlinear expression $f(x) = f_{i,j}(x)$ in Model (5.51), by a newly introduced variable y. Then, we approximate $f(x)$ by a piecewise linear function $\phi(x)$, such that for $\varepsilon > 0$ the approximation satisfies $|f(x) - \phi(x)| \leq \varepsilon$ for all relevant x. Next, we introduce another variable y' for the approximate value of $f(x)$, with $\phi(x) - \varepsilon \leq y' \leq \phi(x) + \varepsilon$, and add the constraint $y = y'$ to our model. Since $f(x) - \varepsilon \leq \phi(x) \leq f(x) + \varepsilon$, any feasible point of the original MINLP is still feasible and we obtain a relaxation.

The resulting piecewise linear approximations are typically still nonconvex. However, any piecewise linear function can be modeled within a mixed-integer linear program by using additional binary variables and linear constraints and thus our relaxation of Problem (5.51) can be formulated as an MIP. An important fact is that the resulting MIPs can be solved comparatively faster than MINLPs in practice; see, e.g., [16, 24].

The drawbacks of the approach described so far are that MIP-based relaxations suffer from the curse of dimensionality and the solution of an MIP relaxation might only satisfy the constraints up to a tolerance of ε. The first disadvantage is due to the fact that the size of the MIP formulations, and especially the number of binary variables, depend on the number of pieces of the piecewise linear functions. In order to compensate for this drawback as much as possible, we are interested in piecewise linear approximations to nonlinear functions with a minimal number of pieces.

To determine a piecewise linear approximation $\phi^*(x)$ of a univariate continuous function $f(x)$ with approximation error ε^* and a minimal number of k^* pieces that

Algorithm 3: Computation of a piecewise linear approximation with error at most ε and a minimum number of pieces

Input: A continuous function $f : [a, b] \to \mathbb{R}$, an upper bound on the overall error $\varepsilon > 0$, and an upper bound on the minimal number of pieces k^+.

Output: An optimal piecewise linear approximation ϕ^* of f on $[a, b]$, i.e., a piecewise linear approximation with overall approximation error $\varepsilon^* \leq \varepsilon$ and a minimum number of pieces k^*.

1 Initialize the lower bound k^- on k^* to $k^- \leftarrow 0$.
2 **while** $k^+ - k^- > 1$ **do**
3   Choose some k with $k^- < k < k^+$.
4   Choose arbitrary points $a = \tilde{x}_0 < \tilde{x}_1 < \cdots < \tilde{x}_k = b$.
5   Compute $\tilde{\varepsilon}_i = \min_{\alpha_i, \beta_i \in \mathbb{R}} \max_{x \in [\tilde{x}_{i-1}, \tilde{x}_i]} |f(x) - (\alpha_i x + \beta_i)|$ for $i = 1, \ldots, k$ and let $\phi_{k,i}(x) = \alpha_i x + \beta_i$ be the solution.
6   **while** $\min_{i=1,\ldots,k} \tilde{\varepsilon}_i < \max_{i=1,\ldots,k} \tilde{\varepsilon}_i$ **do**
7   Compute $\tilde{c}_i = \tilde{\varepsilon}_i / (\tilde{x}_i - \tilde{x}_{i-1})^2$ for $i = 1, \ldots, k$.
8   Solve the system of equations $x_0 = a$, $x_k = b$, $\tilde{c}_i (x_i - x_{i-1})^2 = \varepsilon$, $i = 1, \ldots, k$ for the unknowns $x_0, \ldots, x_k, \varepsilon$. Let $\tilde{x}$ be the solution.
9   Compute $\tilde{\varepsilon}_i = \min_{\alpha_i, \beta_i \in \mathbb{R}} \max_{x \in [\tilde{x}_{i-1}, \tilde{x}_i]} |f(x) - (\alpha_i x + \beta_i)|$ and let $\phi_{k,i}(x) = \alpha_i x + \beta_i$ be the solution for $i = 1, \ldots, k$.
10   **if** $\min_{i=1,\ldots,k} \tilde{\varepsilon}_i > \varepsilon$ **then**
11   Update $k^- \leftarrow k$.
12   **if** $\max_{i=1,\ldots,k} \tilde{\varepsilon}_i \leq \varepsilon$ **then**
13   Update $k^+ \leftarrow k$, $\phi^+ \leftarrow \phi_k$, and $\varepsilon^+ \leftarrow \max_{i=1,\ldots,k} \tilde{\varepsilon}_i$.
14 Set $k^* \leftarrow k^+$, $\phi^* \leftarrow \phi^+$, and $\varepsilon^* \leftarrow \varepsilon^+$.
15 **return** k^*, ϕ^*, and ε^*

satisfies an a priori error tolerance $\varepsilon \geq \varepsilon^* > 0$ we make use of Algorithm 3 proposed in [19]. It can be summarized as follows: In the inner loop a piecewise linear approximation ϕ_k with minimal overall maximum error for a fixed number of k pieces is determined. The key ingredient to solve the optimization problems in Steps 5 and 9 is the so-called minimax approximation theory; see, e.g., [25, 31]. One aim of the minimax theory is to find the unique polynomial of degree at most n (here $n = 1$ to obtain linear approximations) that minimizes the maximum approximation error to a continuous univariate nonlinear function. The most famous algorithm to construct such a minimax approximation and hence a possibility to solve the optimization problems is known as Remez' Algorithm [26]. During the inner loop, the algorithm estimates new breakpoints $a = \tilde{x}_0 \leq \tilde{x}_1 \leq \cdots \leq \tilde{x}_{k-1} \leq \tilde{x}_k = b$ based on an approximation problem and updates them by the exact evaluation of the minimax approximation on the intervals resulting from these breakpoints. The fact that the piecewise linear approximation obtained after termination of the inner loop is the maximum-error minimizer for k pieces is an immediate consequence of the following theorem:

Theorem 4 ([18]). *Let $f : [a, b] \to \mathbb{R}$ be a continuous function. Further let $p = (p_i : [x_{i-1}, x_i] \to \mathbb{R})_{i=1}^{k}$ and $p' = (p'_i : [x'_{i-1}, x'_i] \to \mathbb{R})_{i=1}^{k'}$, with $k' \leq k$ be two piecewise minimax approximations of degree n of f over $[a, b]$ with respective errors $\varepsilon_i := \max_{x \in [x_{i-1}, x_i]} |p_i(x) - f(x)|$ and $\varepsilon'_i := \max_{x \in [x'_{i-1}, x'_i]} |p'_i(x) - f(x)|$. Then,*

$$\min\{\varepsilon_i : i = 1, \ldots, k\} \leq \max\{\varepsilon'_i : i = 1, \ldots, k'\}.$$

The outer loop in Algorithm 3 is a one-dimensional search over the number of pieces k while maintaining a valid bracketing $k^- < k^* \leq k^+$ of the minimum number of pieces k^*. A simple adaptive refinement algorithm could be applied for the purpose of determining the initial upper bound k^+: Start with only two breakpoints and compute the minimax approximation on the corresponding interval. If the approximation error tolerance is not satisfied then insert a new breakpoint, e.g., at the midpoint or the point where the maximum error is attained. Then iteratively recompute the minimax approximation on each piece until the desired error tolerance is satisfied.

Unfortunately, there are no convergence results for this algorithm known yet, but the algorithm works well for the functions appearing in water network models. The following theorem summarizes the main theoretical property of Algorithm 3:

Theorem 5 ([18, 19]). *If Algorithm 3 terminates, it returns a piecewise linear approximation ϕ^* with a minimum number of pieces k^* such that the maximum error over each piece is minimized and is equal to ε^*. Moreover at any time a lower bound k^- and an upper bound k^+ on the minimal number of pieces is provided.*

To model the resulting discontinuous piecewise linear approximations with k line segments—given by a sequence of $2k+1$ breakpoints $(x_0, y_{0,r}), (x_1, y_{1,l}), (x_1, y_{1,r}), \ldots, (x_{k-1}, y_{k-1,l}), (x_{k-1}, y_{k-1,r}), (x_k, y_{k,l})$ and an approximation error ε—in terms of mixed-integer linear constraints, we apply the so-called disaggregated incremental model:

$$x = x_0 + \sum_{i=1}^{k}(x_i - x_{i-1})\delta_i, \tag{5.52a}$$

$$y = y_{0,r} + \sum_{i=1}^{k}(y_{i,l} - y_{i-1,r})\delta_i + \sum_{i=1}^{k-1}(y_{i,r} - y_{i,l})z_i + e, \tag{5.52b}$$

$$z_i \in \{0, 1\}, \quad \delta_{i+1} \leq z_i \leq \delta_i \quad \text{for } i = 1, \ldots, k-1, \tag{5.52c}$$

$$0 \leq \delta_1, \quad \delta_k \leq 1, \quad -\varepsilon \leq e \leq \varepsilon. \tag{5.52d}$$

For a description of this modeling technique, as well as for further details on the theory of piecewise minimax approximations, we refer to [19].

Since all nonlinear expressions appearing in Equations (5.51) are separable, each of them can be written as a composition of univariate function and thus we can apply the above procedure recursively to obtain an MIP relaxation of (5.51).

5.2.3 Computational results

Implementation aspects. The presented model and the described solution approach are implemented in C++ as EWave-DOPT .

In a first step the given network is decomposed into active and passive components. To avoid micro components we force every passive component to contain at least five edges and merge it with the neighboring active components otherwise. This way we circumvent the performance loss resulting from too many ADM iterations with trivial subproblems.

For the MINLP subproblems resulting from active components we implement a generic variable bound tightening step, based on feasibility-based bound tightening (FBBT) [1]. With such a preprocessing we can tremendously reduce the number of segments for the piecewise linear relaxations and hence the required number of auxilary variables. At first, for each nonlinear function a forward and backward bound propagation based on interval arithmetics is performed. Subsequently nonlinearities are replaced by linearized convex under- and concave overestimators; for all nonlinear functions in our model even convex and concave envelopes are available. For the resulting linear relaxation tightened bounds are obtained from the FBBT fixed point by solving the linear program described in [1]. We reapply both steps until no further bound improvements are possible. The final bounds of this procedure are applied to our MINLP and a MIP relaxation as described in Section 7 is constructed. To strengthen this relaxation we add linearized convex and concave envelopes for all nonlinear functions as in the preprocessing step. Finally, we reduce big-M coefficients of the MIP [27] to improve numerical stability and solve it with Gurobi 7.0.1 [14]. Default parameters are used, except for enabling the zero-objective heuristic. A warm start with the solution from the last (P)ADM iteration is performed.

To solve the NLP subproblems optimization procedure. We did not use a global approach, since tests show that it is impossible to solve large passive subnetworks (tens to hundreds of elements) and up to 96 time steps. To perform the simulation we sequentially solve the set of equations for each time step. At each time step we apply newton's method and use the tank's filling levels as initial levels in the next time step. Linear systems of equations are solved with CSparse [5]. Gradients for the optimization are computed from discrete adjoint equations, solved backwards in time. The interior point solver Ipopt 3.12.6 [32] as nonlinear optimization framework. We omit a more detailed description of this procedure, since unless a different model is used this is very similar to the optimization procedure described in the Section 5.3.

Penalty parameters are updated after each ADM iteration and not only after convergence to a partial minimum as discussed in the numerical results sections of [10, 9]. We start with very small penalty parameters and multiply each parameter by a factor of 10 if the corresponding coupling constraint is violated.

Instances description. To describe our method and show some optimization results we use the pilot network from the area Holsterhausen introduced in Chapter 2, particularly Section 2.3. This area consists of water works Dorsten-Holsterhausen and a corresponding distribution network. The network model used for the EWave system, and hence our optimization module, is a result of the aggregation techniques described in Chapter 6. This way tens of thousands of pipes of the GIS-based data are reduced to an abtracted network model with 134 nodes and 168 edges. The node set consists of 8 real and 50 artificial tanks and 76 junctions, including supply and demand nodes. The edge set consists of 100 pipes, 20 pumps (variable and fixed speed), 33 valves, and 15 connections. See Figures 7.1–7.2 for a visual representation of the network and Section 7.2 in general to obtained more details about the pilot network.

In addition we have some generic (i.e., component independent) restrictions, such as flow-based mixing ratios and requirements that specific pumps are not used simultanously or only mutually exclusive.

We use a representative day and optimize a time horizon of 24 hours with a resolution of 15 minutes, resulting in 96 time steps.

All compuations are performed on a Intel Core i7 quad core CPU with up to 3.1 GHz, 16 GB main memory and Ubuntu 14.04 linux operating system.

Results. The network decomposition yields three active components with 45, 5, 5 nodes and 60, 5, 9 edges as well as one passive component consisting of 83 nodes and 94 edges.

The MINLP models of the three active components have 66850, 6340, 10972 variables (7196, 722, 1444 binary) and 87246, 7936, 17708 constraints (7548, 1020, 1632 nonlinear). Bound tightening reduces the feasible domains and ranges of nonlinear functions by roughly 90%. Subsequent big-M coefficient preprocessing yields a reduction of roughly 87%. The resulting MIP relaxations of the MINLPs contain 140204, 9972, 45626 variables (36325, 1518, 16642 binary) and 220252, 20646, 61976 linear constraints.

It took 4 PADM iterations and roughly 13 minutes to solve the instance. The overall running time is mainly dominated by the Gurobi MIP solve of the large active component.

Futher results and details from the field test of the EWave system are given in Chapter 11.

5.3 Continuous optimization

For given/fixed discrete control decisions derivative-based optimization techniques can be used to solve the resulting continuous optimization problem with the original nonlinear (discretized) model equations. This can be done with our software tool Anaconda (see Chapter 4, Section 4.4). The considered optimal control problem is of the form

$$\begin{aligned}
\min_{u} \quad & f(y(u), u) \\
\text{s.t.} \quad & g(y(u), u) \geq 0 \\
& h(y(u), u) = 0 \\
& u_{min} \leq u \leq u_{max}
\end{aligned} \tag{5.53}$$

with state vector y, control vector u, objective function f, inequality constraints g and equality constraints h. As described in Chapter 4, Section 4.4, the state vector results from solving a set of (nonlinear) equations $E(y, u) = 0$ for y. For this reason, the functions f, g and h can be considered as functions solely depending on the control u. In fact, the state variables y are not visible for the optimization tools we have linked to our software, their interface only contains the control variables u. Actually, we use DONLP2 [29, 28] and Ipopt [32]. Evaluations of the objective function and the constraints for a given control u can be done based on the corresponding simulation results. Further, we apply a first-discretize adjoint approach to compute the so-called *reduced gradients*, i.e., derivatives of f, g and h with respect to u.

Following the description in [17], we demonstrate the procedure for the computation of the reduced gradient of the objective function $f(y(u), u)$. First, the linear system of *adjoint equations*

$$\underbrace{\left(\frac{\partial}{\partial y} E(y(u), u)\right)^T}_{\text{independent of } f} \xi = -\left(\frac{\partial}{\partial y} f(y(u), u)\right)^T \tag{5.54}$$

has to be solved. It is important to notice that the matrix $\frac{\partial}{\partial y} E(y(u), u)$ in (5.54) is independent of the considered function f. Therefore, this matrix and any decomposition of it computed for solving (5.54) only needs to be computed once. After having solved (5.54), we directly get the reduced gradient from

$$\frac{d}{du} f(y(u), u) = \frac{\partial}{\partial u} f(y(u), u) + \xi^T \underbrace{\frac{\partial}{\partial u} E(y(u), u)}_{\text{independent of } f} . \tag{5.55}$$

Here, the matrix $\frac{\partial}{\partial u} E(y(u), u)$ is independent of f and therefore only needs to be computed once, independent of the number of computed gradients.

Since we consider a time-dependent problem, the discretized model equations $E(y, u)$ have a special structure and therewith the linear system (5.54),

$$\frac{\partial}{\partial y} E(y, u) = \begin{pmatrix} I & & & \\ A_1 & B_1 & & \\ & A_2 & B_2 & \\ & & \ddots & \ddots & \\ & & & A_N & B_N \end{pmatrix} \tag{5.56}$$

with

$$A_j = \frac{\partial}{\partial y_{old}} F(t_{j-1}, t_j, y(t_{j-1}), y(t_j), u(t_{j-1}), u(t_j)) \tag{5.57}$$

and

$$B_j = \frac{\partial}{\partial y_{new}} F(t_{j-1}, t_j, y(t_{j-1}), y(t_j), u(t_{j-1}), u(t_j)). \tag{5.58}$$

Thus, for the objective function f, the set of adjoint equations (5.54) reads

$$\begin{pmatrix} I & A_1^T & & & \\ & B_1^T & A_2^T & & \\ & & B_2^T & \ddots & \\ & & & \ddots & A_N^T \\ & & & & B_N^T \end{pmatrix} \begin{pmatrix} \xi(t_0) \\ \xi(t_1) \\ \ldots \\ \ldots \\ \xi(t_N) \end{pmatrix} = - \begin{pmatrix} \frac{\partial}{\partial y_0} f(y,u)^T \\ \frac{\partial}{\partial y_1} f(y,u)^T \\ \ldots \\ \ldots \\ \frac{\partial}{\partial y_N} f(y,u)^T \end{pmatrix}. \tag{5.59}$$

Here, the partial derivatives $\frac{\partial}{\partial y_j}$ refer to the blockwise partitioning of the state vector according to the time steps. Obviously, (5.59) can be solved blockwise (backwards in time), reducing the size of the single systems to be solved.

Bibliography

[1] P. Belotti, S. Cafieri, J. Lee, and L. Liberti, Feasibility-Based Bounds Tightening via Fixed Points. In *Combinatorial Optimization and Applications: 4th International Conference, CO-COA 2010, Kailua-Kona, HI, USA, December 18–20, 2010*, Proceedings, Part I, (W. Wu and O. Daescu, eds.), Springer, Berlin, Heidelberg, 2010, 65–76. https://doi.org/10.1007/978-3-642-17458-2_7.

[2] S. Boyd, N. Parikh, E. Chu, B. Peleato, and J. Eckstein, Distributed Optimization and Statistical Learning via the Alternating Direction Method of Multipliers. *Found. Trends Mach. Learn.* 3(1) (2011), 1–122.

[3] J. Burgschweiger, B. Gnädig, M. C. Steinbach, Optimization models for operative planning in drinking water networks. *Optimization and Engineering* 10(1) (2009), 43–73. http://dx.doi.org/10.1007/s11081-008-9040-8.

[4] C. F. Colebrook, Turbulent flow in pipes with particular reference to the transition region between smooth and rough pipe laws. *JICENG* 11(4) (1939), 133–156.

[5] T. Davis, *Direct Methods for Sparse Linear Systems*, SIAM, 2006.

[6] D. Gabay and B. Mercier, A dual algorithm for the solution of nonlinear variational problems via finite element approximation. textitComputers & Mathematics with Applications 2(1) (1976), 17–40.

[7] B. Geißler, *Towards Globally Optimal Solutions for MINLPs by Discretization Techniques with Applications in Gas Network Optimization*, PhD thesis. Friedrich-Alexander-Universität Erlangen-Nürnberg (FAU), 2011.

[8] B. Geißler, A. Martin, A. Morsi, and L. Schewe, Using Piecewise Linear Functions for Solving MINLPs. In *Mixed Integer Nonlinear Programming* (J. Lee and S. Leyffer, eds.), Vol. 154 The IMA Volumes in Mathematics and its Applications, Springer, New York, 212, 287–314.

[9] B. Geißler, A. Morsi, L. Schewe, and M. Schmidt, Highly Detailed Gas Transport MINLPs: Block Separability and Penalty Alternating Direction Methods. Accepted for publication in *INFORMS Journal on Computing*. DOI: 10.1287/ijoc.2017.0780 (2017).

[10] B. Geißler, A. Morsi, L. Schewe, and M. Schmidt, Penalty Alternating Direction Methods for Mixed-Integer Optimization: A New View on Feasibility Pumps. *SIAM Journal on Optimization* 27(3) (2017), 1611–1636. https://doi.org/10.1137/16M1069687.

[11] B. Geißler, A. Morsi, L. Schewe, and M. Schmidt, Solving power-constrained gas transportation problems using an MIP-based alternating direction method. *Computers & Chemical Engineering* 82 (2015), 303–317.

[12] R. Glowinski and A. Marroco, Sur l'approximation, par éléments finis d'ordre un, et la résolution, par pénalisation-dualité d'une classe de problèmes de Dirichlet non linéaires. *ESAIM: Mathematical Modelling and Numerical Analysis – Modélisation Mathématique et Analyse Numérique* 9(R2) (1975), 41–76. http://eudml.org/doc/193269.

[13] J. Gorski, F. Pfeuffer, and K. Klamroth, Biconvex sets and optimization with biconvex functions: a survey and extensions. *Mathematical Methods of Operations Research* 66(3) (2007), 373–407.

[14] Z. Gu, E. Rothberg, and R. Bixby, *Gurobi Optimizer Reference Manual, Version 5.0*, Gurobi Optimization Inc., Houston, USA, 2016.

[15] S.-P. Han and O. L. Mangasarian, Exact penalty functions in nonlinear programming. *Mathematical Programming* 17(1) 17(1) (1979), 251–269.

[16] T. Koch, B. Hiller, M. E. Pfetsch, and L. Schewe, *Evaluating Gas Network Capacities*, SIAM-MOS series on Optimization, SIAM, 2015, xvii + 364.

[17] O. Kolb, *Simulation and Optimization of Gas and Water Supply Networks*, Verlag Dr. Hut, München, 2011.

[18] C. L. Lawson, Characteristic properties of the segmented rational minimax approximation problem. *Numerische Mathematik* 6 (1964), 293–301. year = 1964

[19] A. Morsi, *Solving MINLPs on Loosely-Coupled Networks with Applications in Water and Gas Network Optimization*, PhD thesis. Friedrich-Alexander-Universität Erlangen-Nürnberg (FAU), 2013.

[20] J. Nikuradse, *Laws of Flow in Rough Pipes*, Technical Memorandum 1292. National Advisory Committee for Aeronautics Washington, 1950.

[21] J. Nikuradse, *Strömungsgesetze in rauhen Rohren*, Forschungsheft auf dem Gebiete des Ingenieurwesens, VDI-Verlag, Düsseldorf, 1933.

[22] J. Nocedal and S. J. Wright, *Numerical Optimization*, 2nd edn., Springer Series in Operations Research and Financial Engineering, Springer, New York, 2006.

[23] I. Nowak, *Relaxation and decomposition methods for mixed integer nonlinear programming*, 2005. urn:nbn:de:kobv:11-10038479.

[24] M. E. Pfetsch, A. Fügenschuh, B. Geißler, N. Geißler, R. Gollmer, B. Hiller, J. Humpola, T. Koch, T. Lehmann, A. Martin, A. Morsi, J. Rövekamp, L. Schewe, M. Schmidt, R. Schultz, R. Schwarz, J. Schweiger, C. Stangl, M. C. Steinbach, S. Vigerske, and B. M. Willert, Validation of nominations in gas network optimization: models, methods, and solutions. *Optimization Methods and Software* 30(1) (2015), 15–53.

[25] M. J. D. Powell, *Approximation Theory and Methods*, Cambridge University Press, 1981.

[26] E. Remez, Sur le calcul effectif des polynomes d'approximation de Tschebyscheff. *C. R. Acad. Sci.* 199 (1934), 337–340.

[27] M. W. P. Savelsbergh, Preprocessing and Probing Techniques for Mixed Integer Programming Problems. *ORSA Journal on Computing* 6(4) (1994), 445–454. https://doi.org/10.1287/ijoc.6.4.445.

[28] P. Spellucci, A New Technique for Inconsistent QP Problems in the SQP Method. *Mathematical Methods of Operations Research* 47(3) (1998), 355–400.

[29] P. Spellucci, An SQP Method For General Nonlinear Programs Using Only Equality Constrained Subproblems. *Mathematical Programming* 82(3) (1998), 413–448.

[30] P. Swamee and A. K. Jain, Explicit eqations for pipe-flow problems. *ASCE J. Hydraul. Div.* 102 (1976), 657–664.

[31] N. Trefethen, *Approximation Theory and Approximation Practice*, SIAM, 2013.

[32] A. Wächter and L. T. Biegler, On the Implementation of a Primal-Dual Interior Point Filter Line Search Algorithm for Large-Scale Nonlinear Programming. *MATHP* 106(1) (2006), 25–57.

[33] R. E. Wendell, and A. P. Hurter Jr., Minimization of a non-separable objective function subject to disjoint constraints. *Operations Research* 24(4) (1976), 643–657.

Part III

Practical aspects

The system architecture for the new decision support system (DSS) follows a modular structure. This is built on the existing SIWA water management systems from Siemens AG. The need for modularity arises, on the one hand, from the demand for transferability to other supply networks similar to the pilot network, which demands that the software modules are independent of the specific network. On the other hand, the DSS permits a host of different applications (e.g. simulation preview, assessment of planning alternatives, optimized plant management), which demand flexible interaction of exchangeable algorithms, models, pre- and post-processing of data (e.g. KPI calculation), as well as an adapted presentation of the input options and results. The involved data management of the input data (static plant data and dynamic operating data) and result data (simulation and optimization results) has a key role. The software architecture of the new tool permits effective, non-volatile data storage and correct, automatic further processing of all this data – regardless of the plant and the application scenario on which it is based. This is guaranteed by the definition of modular exchange formats and suitable interfaces on the basis of current software technologies. Using the architecture outlined here, it is possible to free the user of the DSS from the need for expert mathematical knowledge about complex numerical algorithms and their interaction. The user initiates a required application scenario on the user interface and the assistance system orchestrates the required functionalities and data processing and delivers the corresponding KPIs (key performance indicator) back to the user interface.

Chapter 6

Network aggregation

Tim Jax

Abstract. Simulating real network often proves to be challenging due to the huge number of elements given. For this reason, aggregation strategies are recommended in order to reduce a network's dimensions and, thus, computational efforts. This section introduces corresponding strategies used to create appropriate network models with respect to the EWave project. In this context, a novel approach will be discussed that – particularly based on manual processes – realizes network aggregations with a self-defined extent of pipe reduction and layout flexibility.

6.1 Introduction

Simulating pressure zones requires to consider water works for water treatment, networks for water distribution as well as customers for water consumption. In this context, especially networks consist of multiplicity components, including pumps, valves, tanks and pipes. To reproduce given characteristics properly, influence of these elements on aspects such as flow behavior, pressure distribution and energetic consumption must be taken into account. However, dimensions of real networks are in general too extensive for simulating each of its components in detail. For instance, pressure zones considered as prototype region for realizing the EWave project count up to 15,000 pipe elements.

Implementing such network dimensions completely negatively affects computational effort, even when using high-performance computers. In particular, when having to simulate systems that are based on operating in a short period of time correspondingly increased computing times are a major issue. For example, in the EWave project data is assumed to be refreshed in cycles of about 30 minutes. Hence, the period of computations must be restricted to less than this time horizon. As a consequence, strategies for minimizing network complexity are required in order to reduce the time span for calculating resulting flow conditions.

Handling complex network structures is a common problem in the field of simulating flowing fluids, especially regarding systems of water supply and sewerage. For that reason, there are several publications given in literature that deal with strategies for decreasing network complexity in order to reduce computational efforts while trying to preserve the original network's hydraulic characteristics. Corresponding techniques are usually based upon aggregating methods. As described in [5, 8], some of the first strategies with respect to network aggregation were considered in the end of the 1980s by Hamberg and Shamir in [2] and [3]. In [2] a combining approach for single pipe elements of a system is analyzed and in [3] a continuum approach for all pipe elements of a network is applied. In the mid of the 1990s, Anderson and Al-Jamal presented a parameter-fitting approach in [1]. Here, simplifying given networks is based on defining layouts of the aggregated model a priory. In [9] Ulanicki et al. regarded network aggregation by eliminating system components that were determined by analytical processes. Nearly twenty years later, Perelman et al. describe an approach based on the works by Ulanicki at al. in [5] and [6] that allows to preserve aspects of water quality besides aspects of hydraulics when realizing aggregated network models. In [7] Perelman et al. consider strategies for network simplification using a clustering approach. Reference [7] can be recommended for a bit more detailed literature review, also. Finally, a more recent approach dealing with network aggregation based on Ulianicki et al. is given in [4].

Nevertheless, most strategies introduced in literature prove to be less appropriate regarding requirements given with respect to network aggregation in the EWave project. Focusing primarily on a significant reduction of computing times, the EWave project requests applying an approach that enables to decrease complexity of any real network region by more than 90% in a user-friendly and flexible way. So, the approach must allow for realizing very abstract aggregated network models that can be derived directly from a given real topology. Besides, the layout of the aggregated network model should be controllable.

Most approaches for network aggregation defined in literature are based on fully automatic generation including predefined search algorithms. Because they are steered completely by the program given, they offer just restricted possibilities to control the final pipe number and the final layout design. Hence, to apply network aggregations that reduce the number of pipes but also ensure the flexibility of users with respect to a self-defined extend as intended, a novel approach is required for the EWave project that – contrary to strategies given in literature – is much more based on manual processes.

In this chapter theory and realization of the strategy developed and applied within the EWave project for reducing dimensions of any given water supply network are introduced. After layout definition by choosing relevant paths and nodes for the final network model, the approach is characterized by an iterative process of step-wise pipe reduction via neglecting and combining given elements repeatedly. Neglected pipes and water demands are finally taken into account by artificial tank elements and artificial sinks. The strategy allows for implementing quite abstract network models whose number of pipe elements is reduced to a desired minimum but still allows to preserve hydraulic conditions of flow and pressure in predefined regions after including appropriate steps of calibration. Also, the software-tool TWaveGen is presented that supplements the TWaveSim environment introduced in Chapter 4, Section 4.3 by realizing the novel approach for network aggregation within a MATLAB graphical user interface.

Subsequent sections are organized as follows: Section 6.2 deals with theoretical aspects. It details objectives and steps of the network aggregation realized for the

EWave project. Section 6.3 presents implementation in MATLAB by introducing the software-tool TWaveGen. It describes the graphical user interface developed, relevant algorithms and features incorporated. Finally, Section 6.4 gives a conclusion and outlook. Continuative contents such as applying the strategy for network aggregation to a prototype test region as well as steps for calibration are given in Chapter 7, in particular Sections 7.2 and 7.3.

6.2 Theoretical aspects

6.2.1 General objectives

As indicated in the introduction, an adequate strategy enables to decrease effort of network computation by reducing the number of pipe elements while ensuring the flexibility of model design. So, an approach is required that allows to realize a reasonable compromise of both reduced complexity and extended flexibility when creating a model. In this context, most applications for network aggregation introduced in literature concentrate on preserving overall network characteristics with respect to hydraulics and minimizing a user's amount of work when realizing corresponding simplified models. As a consequence, they may conflict the general objective of most extended pipe reduction and most extended layout flexibility. For example, computational effort can be reduced more effectively if hydraulic characteristics were preserved just within a small number of specific regions as it allows to neglect more pipe elements. Analogously, trying to minimize a user's amount of work when realizing simplified models usually leads to full-automatic approaches based on predefined search algorithms. Although most of these implementations allow to identify special points within the original network that should be involved in the resulting model, they offer no possibilities to control the paths for connecting the points defined. So, the user is obliged to rely on results of search algorithms embedded. As a consequence, resulting paths might be appropriate connections from the algorithmic point of view, but they might be less consistent with relevant paths known by staff members, thus leading to a model whose layout could be far from the original network. The only way of counteracting this disadvantage would be to include additional points. However, this would increase the number of pipe elements and, thus, affect computational effort negatively.

To overcome these restrictions, general purpose of the modeling approach realized for the EWave project is to reduce computational effort by novel strategies that allow for the largest self-defined pipe reduction and layout flexibility possible. This generally demands for implementing a manual process. However, a manual process increases the effort of aggregation compared to full-automatic solutions. A corresponding classification of EWave compared with common full-automatic approaches described in literature is visualized in Figure 6.1.

In the following, the aspects of pipe reduction, layout flexibility and manual procedure are detailed.

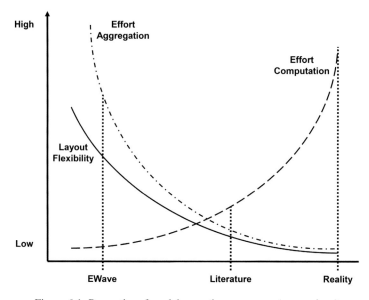

Figure 6.1. Properties of model genrating processes (comparison)

Reducing pipes. Reducing computational efforts demands reducing network dimensions significantly. In this context, the number of pipes remaining in a network model is usually determined by implemented strategies to realize the aggregation process. Regarding applications of full-automatic aggregation algorithms, this generally leads to network models that require to compute hundreds of remaining pipes still. However, simulating flow behavior within hundreds of pipes is too expensive when having to satisfy the strict time targeting considered in the EWave project. Therefore, an approach for aggregating real network topologies is required that enables a significant reduction of elements down to a predefined number of remaining pipes that is independent of the aggregation algorithms implemented. This will allow for extremely abstract network models that probably will not be able to preserve hydraulic conditions in every of its sections. But focusing primarily on aspects of significantly reducing computational efforts than on preserving hydraulic conditions, corresponding abstract network models are about to be accepted that do not deliver physically precise information in each region. The idea is to implement strategies that preserve hydraulic conditions at least within a few number of predefined regions by including adequate calibration techniques that will be further detailed in Chapter 7, Section 7.3.2. In fact, implementations of a corresponding aggregation process would be able to realize theoretical network reductions of more than 99% and mark a significant contrast to common aggregation techniques defined in literature whose number of remaining pipes is generally determined by involved algorithms.

Extending flexibility. Besides reducing pipe elements, the approach for network aggregation should ensure an extended flexibility regarding the layout design of finally resulting network models. This objective marks another contrast to aggregation strategies presented in literature where the final design is generally determined by given automatic algorithms. Hence, novel strategies are required that allow for extended flexibility in the sense of granting full control with respect to the aggregation process. Finally, this benefits two relevant aspects: First, the implemented approach is independent of special network characteristics and capable of including them in the resulting model. For example, locations can be taken into account where flow conditions are affected or measured. Second, the implemented approach will be able to include novel features compared to known processes presented in literature. For example, paths to connect relevant points can be determined directly and combinations of network regions using different grades of aggregation can be realized (i.e. combining abstract regions with extensive aggregation besides detailed regions with minor aggregation for the purpose of detailed information in specific regions or realizing comparative computations). Moreover, full control allows to implement network models that – despite a significant reduction of given elements – are able to preserve the general look of the original network. In the end, these features help to benefit acceptance of resulting network models by staff members and to involve their expertise in the process of network generation directly.

Manual approach. Combining significant pipe reduction and extensive layout flexibility can hardly be realized by fully automated processes presented in literature. Indeed, realization by a manual approach seems to be more effective. However, manual aggregation drastically increases the effort of model generation and, thus, corresponds to a drawback compared to common strategies defined in literature as visualized in Figure 6.1. For this reason, strategies for network aggregation need to be realized in a user-friendly way that reduces efforts of manual network generation to a reasonable degree, especially regarding the overall amount of working hours when having to deal with network topologies that consist of thousands of pipe elements. For this purpose, despite being a manual approach primarily, as much steps as possible should be supplemented by some automatic support. In this context, the term user-friendly also includes that the approach of network generation can be applied by anyone without intensive training. So, the overall process of network aggregation should be straightforward and its single steps should be comprehensible.

6.2.2 General concept

The concept for generating aggregated network models in the EWave project is based on steps visualized in Figure 6.2. They allow to decrease computational efforts by reducing the number of given pipes in three steps: First, defining a model layout by neglecting elements given for real network topologies. Second, reducing the number of pipes within this model by combining and further neglecting elements. Third, collecting information of neglected elements within artificial tanks and including drinking water demands by defining artificial sinks connected to these tanks.

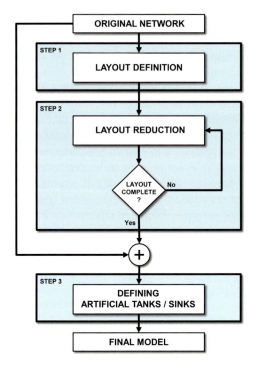

Figure 6.2. Steps for generating a network model

Each of these steps is realized by manual techniques in order to ensure flexibility during the generation procedure. In this context, especially reducing elements of the model layout resulting from the first step corresponds to an iterative process. This means, a preliminary state of network simplification corresponds to the starting point for realizing another, subsequent state of simplification. As a consequence, different levels of aggregation can be directly realized and used (e.g. for comparing purposes). Because there are no limits regarding the resulting grade of simplification within this concept, the whole network could be theoretically reduced down to any desired amount of remaining elements.

The idea behind the general concept visualized by steps in Figure 6.2 is as follows: Each pipe is spatially discretized by appropriate discretization techniques defined in Chapter 4, Section 4.3.2. As a consequence, computational effort in simulation is mainly determined by the number of pipes and discretization points considered. Hence, the less pipes are included, the faster will be their computation. However, in order to preserve information of pipes given within the original network topology, especially regarding overall mass and demand of drinking water, it should not be completely neglected when reducing the number of elements during the generation process. For this reason, information is preserved by defining artificial tanks and artificial sinks that are computed using equations detailed in Chapter 4, Section 4.2.2. This way, total amount of water and demands of the original network is still considered without having to include additional spatial discretization. The final model will then consist of just a reduced number of pipes besides several artificial tanks and a couple of elements really present within the original network such as pumps and tanks.

Within the following subsections, assumptions, requirements and processes for realizing steps visualized in Figure 6.2 are described in more detail.

6.2.3 Assumptions and requirements

Implementing and applying the procedure introduced requires information of the real network topology that is about to be aggregated. This information must be given by responsible and experienced staff members of the water supplier. In this context, detailed properties are needed, including position and function of contributing waterworks as well as characteristics regarding relevant elements such as pumps, valves and tanks within the network. Corresponding plants are usually responsible for regularizing pressure or storing water and, thus, should be involved within the final network model. Also important is information regarding the distribution of pressure zones together with details on their mutual interaction. In addition, knowing about the localization of industrial consumers and particular paths (such as special-purpose pipes) that are assumed to have a considerable influence on flow behavior is relevant for generating a suitable simulation model.

Besides general network characteristics, details with respect to given pipes must be provided. All pipes are assumed to be straight elements. In this context, Gauss-Krueger coordinates of a pipe's origin and end as well as information about its diameter and length are required (especially as for originally curved elements correct lengths cannot be derived by knowing the coordinates of a pipe's origin and end alone). Distinct identification numbers must be assigned to each pipe as well as its points of origin and end. These identification numbers are needed for implementing the network aggregation by a manual approach properly as indicated in Subsection 6.3.2 and Subsection 6.3.3.

Furthermore, an affiliation of each pipe to given pressure zones and supply zones of existent waterworks is recommended. Corresponding details are helpful for realizing and calibrating artificial tank elements as described in Chapter 7, Section 7.3.3. Additional information such as material and friction parameters can be also useful. However, regarding thousands of elements in reality, such information often proves to be incomplete. For this reason, especially friction values are considered to be additional calibration parameters that can be altered in order to reproduce hydraulic conditions within the final model more accurately. Having an affiliation of each pipe to given streets and cities is optional, but it can be useful to determine relevant interactions of pipes due to given connections and to identify appropriate paths with respect to the process of layout definition. Finally, drinking water demands for each pipe must be known in order to realize sinks connected to artificial tanks whose realization is detailed in Subsection 6.2.6. Further consumer information is also optional but could be effectively used in the process of defining a layout, too.

Another information that must be provided is data with respect to given hydraulic conditions. In this context, localization of at least a few points within the network should be known that represent the position of instruments to measure pressure distribution and flow velocity. Such points are required for calibration processes and used to compare simulation results with real conditions. The position of these points must be known a priori in order to consider them when aggregating the network. In this context, when implementing very abstract network versions the model reduction might be too large so that there is little chance of preserving hydraulic conditions, even when applying calibration strategies. For this reason, particularly measurements nearby the outlets of waterworks are assumed to be relevant for calibration since the aggregation level generally proves to be little here.

A clearly arranged list including most relevant and some optional data for implementing the manual approach of network generation is given in Table 6.1. In this context, missing information with respect to relevant data listed does not mean that the approach for network generation presented cannot be applied at all. However, a lack of corresponding data impedes generating appropriate network models.

6.2.4 Layout definition

First step in the network aggregating process according to Figure 6.2 is to determine an appropriate layout. The layout considers all paths that should contribute to the final network model. This means, ways on which water is transported from waterworks through given pressure zones are pre-defined. Identifying a layout corresponds to a selection of pipes within the original network topology that are assumed to be specifically relevant for water transport. Other elements given are temporarily neglected. For this reason, deciding for appropriate layout structures should be based on discussions that include the expertise of experienced waterworks staff members that are familiar with the network and its characteristics. In general, important factors that may determine the design of first model layouts are given by aspects such as:

- **Realizing structures comparable with the original network.**
  In order to benefit acceptance of the final simulation model and its results by waterworks staff members, it is assumed to be advantageous realizing a layout that is most comparable to structures of the original network.

- **Including paths having an utmost meaning for flow behavior.**
  Among special paths with utmost meaning for flow behavior are outlets of waterworks or special-purpose pipes. Outlets of waterworks play a key role to distribute water into correct directions and are relevant for calibration. Special-purpose pipes are in general responsible for transporting water over far distances without having numerous branches into sub-regions.

- **Considering connections to industrial customers.**
  Industrial consumers are characterized by significant drinking water demands and thus influence flow behavior within the whole network. Hence, including localizations of industrial consumers helps to reproduce characteristics of flow more realistically.

Data with respect to the general network		
Information	**Relevance**	**Required for**
Position / Characteristics Waterworks	relevant	Generation
Position / Characteristics Tanks	relevant	Generation
Position / Characteristics Pumps	relevant	Generation
Position / Characteristics Valves	relevant	Generation
Position Measuring Points	relevant	Generation / Calibration
Position Industrial Consumers	relevant	Generation / Calibration
Distribution Special-Purpose Pipes	relevant	Generation
Distribution Pressure Zones	relevant	Generation / Calibration
Distribution Supply Zones	relevant	Generation / Calibration

Data with respect to pipe elements		
Information	**Relevance**	**Required for**
Identification Number (ID)	relevant	Generation
Length	relevant	Generation
Diameter	relevant	Generation
GK-coordinates initial point	relevant	Generation
ID initial point	relevant	Generation
GK-coordinates end point	relevant	Generation
ID end point	relevant	Generation
Material	optional	Calibration
Friction Parameter	optional	Calibration
Demand	relevant	Generation / Calibration
Consumer	optional	Generation
Pressure Zone	relevant	Generation / Calibration
Supply Zone	optional	Generation / Calibration
Borough	optional	Generation
Street	optional	Generation

Table 6.1. Relevant and optional data for the network aggregation

- **Involving representatives for pipe distributions in major cities.**
  Similar to industrial consumers concentrated agglomerations of pipes in major cities are characterized by significant drinking water demands that influence flow behavior within the whole network. Thus, including paths and crossroads that represent the localization of major cities in a simplified and reduced way might also help to reproduce flow characteristics more realistically.

For choosing appropriate paths given information with respect to diameters and lengths can be helpful. In water supply just pressurized pipes are considered. So, diameters and lengths are indicators for the amount of water transported within special regions. Hence, it can be reasonable to select paths given by connected pipes that are characterized by large diameters and lengths as they are responsible for transporting huge amounts of water. Also, the affiliation of pipes to pressure zones, boroughs and streets can be used to identify paths that are suitable for simulating water transport as they help to evaluate connections among elements given. Besides, paths that involve relevant localizations should be included in the layout definition. Relevant localizations are especially given by:

- Positions of waterworks.
- Positions of pumps, valves and tanks.
- Positions of measuring instruments.
- Positions of consumers with significant demand.

As the process of layout design in the EWave project is given by a manual approach, the resulting model layout can be entirely determined by the user and thus allows to include relevant paths and points without conflict. Because the layout definition reduces a complete network consisting of thousands of elements to a first collection of special paths, it generally corresponds to the step with the most significant pipe reduction in the overall aggregation procedure visualized by Figure 6.2. However, as it is based on manual processes, defining a layout also corresponds to the step that costs the most time regarding the complete aggregation procedure. In this context, the total amount of working hours significantly depends on preliminary studies and preparations with respect to the question how the resulting layout should look like. In general, the approach is performed much faster when it is realized by experienced staff members. Besides, though being manual it can be sped up by appropriate algorithms as described in Section 6.3.

A real-life example regarding layout definition with respect to the network considered in the EWave project can be found in Chapter 7, Section 7.2.2.

6.2.5 Pipe aggregation

A given layout corresponds to a first level of aggregation. However, the corresponding model might consist of hundreds of remaining pipes still. So, in order to further reduce the amount of elements within this model, the layout is used as a starting point for decreasing the number of pipes by an iterative process. The procedure is primarily based on combining remaining pipes but also allows for neglecting elements. In this context, this subsection focuses on the aspect of combining pipes. When combining pipes, a special number of elements is manually chosen and afterwards united to a single straight element with modified characteristics that connects the outer points of the first and last elements selected as visualized in Figure 6.3.

Major advantage of combining pipes is that the original discretization applied to each pipe according Chapter 4, Section 4.3.2 now must be applied just to one single element. Assuming a standardized discretization of $nx_i = 100$ cells per pipe i, the example visualized in Figure 6.3 thus safes up to 600 differential equations when combining the four pipes given to the single element shown. This is because three pipes discretized by 100 cells can be saved and for each cell two differential equations would have been computed, one for the conservation of mass and one for the conservation of momentum. Hence, this procedure allows for some significant reduction of computational efforts, especially when combining given pipe elements repeatedly by iteration.

In order to preserve original information, particularly the total mass of water within pipes combined, the resulting element must be characterized by adapted parameters with respect to length, diameter and friction. In the following, let n denote the total number of elements chosen to define a new pipe. Given an individual length l_i for each pipe selected, the resulting length L of the combined element simply equals their summation. Thus, it holds:

$$L = \sum_{i=1}^{n} l_i.$$

As indicated by Figure 6.3, resulting length L does not correspond to the real length of a straight line connecting the points of origin and end that remain after combination. However, due to the data structure used with respect to the EWave system, a length for computing flow conditions that differs from the real length between points of origin and end can be assigned to each pipe element without conflict.

When all elements selected for a combination are also characterized by different diameters d_i the resulting diameter D of the combined element must be computed by using an appropriate mean value. For this purpose, volumes v_i resulting for each pipe selected must be determined first and summed up to a total volume V that has to be preserved by the combining element. This yields:

$$V = \sum_{i=1}^{n} v_i \quad \text{with} \quad v_i = \pi \left(\frac{d_i}{2}\right)^2 l_i.$$

Diameter D of the combined pipe element can then be determined based on parameters L and V just evaluated, leading to:

$$D = 2\sqrt{\frac{V}{\pi L}}.$$

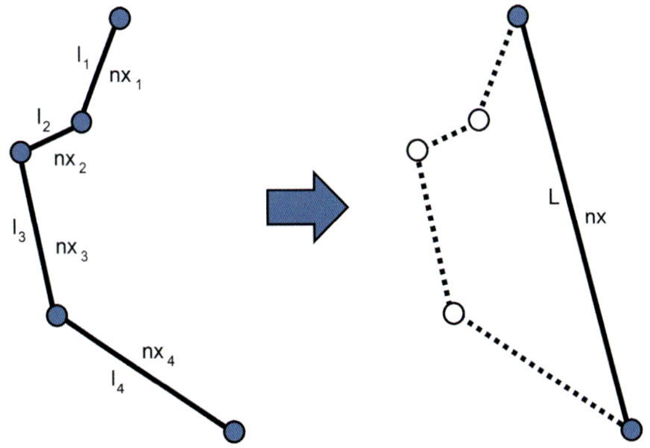

Figure 6.3. Combining pipes

Finally, pipes selected for combination might also be characterized by varying friction parameters. In fact, as mentioned in Subsection 6.2.3 before, friction values primarily correspond to calibration parameters and thus can be altered. Nevertheless, assuming different friction parameters k_i of the pipes that should be unified, a first friction value K of the combining element is determined by forming an appropriate mean value that considers a weight based on known pipe lengths. It holds:

$$K = \frac{1}{nL} \sum_{i=1}^{n} l_i k_i.$$

The process of combining pipes can be repeatedly applied to already combined elements. Thus, this approach leads to network models characterized by a reduced number of pipes that can be used as starting point for further pipe reductions. This way a final network model that considers a predefined number of remaining elements can be determined by an iterative procedure. Theoretically, it is possible to apply this approach until just a drastically reduced number of one single combining element remains. However, the abstraction of corresponding models would be immense and increases the risks of negatively affected hydraulic conditions and acceptance.

6.2.6 Generating artificial tanks

Based on the original network topology and the resulting network model that is given after defining a layout and repeatedly combining pipes to reduce the overall amount of elements, information of all pipes neglected during the aggregation process will be considered by a number of artificial tanks. Artificial tanks are primarily used to preserve the total mass given within the original network. This means, volumes of drinking water within these artificial tanks correspond to volumes of drinking water within real pipes neglected. The purpose of regarding tanks instead of pipes is a significant reduction of equations to solve during simulation as tank elements avoid having to consider a spatial discretization. Hence, overall mass within the system is preserved while computational efforts are kept small.

In order to achieve reasonable results by regarding artificial tanks their parameters must be well defined. As described in Chapter 4, Section 4.2.2, tank elements correspond to nodes whose state variable $H_T(t)$ (i.e. pressure head within a tank) is characterized by:

$$H'_T(t) = \frac{1}{A_{Tank}} (Q_T(t) + Q_S(t)), \qquad H_T(t_0) = H_0. \qquad (6.1)$$

Here, A_{Tank} denotes a constant base area, $Q_S(t)$ represents an external inflow or outflow and $Q_T(t)$ is the flow given by connected edges. Assuming the tank to be connected to the outlet of an element j and the inlet of an element k, it follows $Q_T(t) = Q_R^j(t) - Q_L^k(t)$ with $Q_R^j(t)$ being the flow at the outlet (subscript R = right) of element j and $Q_L^k(t)$ being the flow at the inlet (subscript L = left) of element k. Coupling conditions can be expressed via the pressure loss that is caused by each inflow and outflow using:

$$\tilde{\zeta} Q_R^j(t)|Q_R^j(t)| = H_R^j(t) - H_T(t),$$
$$\tilde{\zeta} Q_L^k(t)|Q_L^k(t)| = -\left(H_L^k(t) - H_T(t)\right).$$

Parameters A_{Tank} and $\tilde{\zeta}$ correspond to relevant values that must be appropriately dimensioned in order to get artificial tanks that are able to replace neglected pipe elements appropriately. In this context, $\tilde{\zeta}$ is considered to be a calibration parameter that can be altered to improve simulation results.

An artificial tank element generated usually represents a special subset of neglected pipes. For determining appropriate values of A_{Tank}, let the total volume within these neglected pipes be denoted by V and the sum over all their lengths by L. Now, assuming all these neglected pipes to be combined to a single element, conservation of mass within this resulting pipe can be described via

$$H_t'(t) + \frac{a^2}{gA_{Pipe}L}(Q_R(t) - Q_L(t)) = 0 \tag{6.2}$$

when regarding mass conservation defined by

$$\frac{\partial h}{\partial t} + \frac{a^2}{gA_{Pipe}}\frac{\partial q}{\partial x} = 0 \tag{6.3}$$

and discretization by a single cell. Here, a denotes sound velocity, g describes gravitational constant and $A_{Pipe} = V/L$ represents the pipe cross-section area. By comparing resulting formulas for conservation of mass within the artificial tank defined by (6.1) and within the representative pipe defined by (6.2) equivalence is given when using

$$A_{Tank} = \frac{gV}{a^2}. \tag{6.4}$$

Values of A_{Tank} derived by this formula correspond to an appropriate default value for for dimensioning artificial tanks. However, in order to improve simulations, values of A_{Tank} can be altered regarding calibration processes. In this case, manipulated values must be also considered vice versa with respect to resulting water elevation levels given within the artificial elements.

Like previous steps of network aggregation described, defining artificial tanks is a manual approach. In this context, data of the original network topology and data of the generated network model must be considered in order to identify all pipes that were neglected during the aggregation process. Knowing the pipes neglected, special subsets of these elements can be selected in order to compute a representing artificial tank element. The amount of pipes collected within such tanks as well as nodes for connecting them with the network model are manually defined. However, the more neglected pipes are collected within a reduced number of artificial tanks, the larger is the resulting level of abstraction and thus might negatively affect hydraulic conditions. In order to preserve hydraulic conditions at least within special regions of the resulting network model appropriate calibration techniques detailed in Chapter 7, Section 7.3.3 are required.

6.2.7 Sink realization

Steps described in previous subsections concentrate on the general approach for aggregating a given network topology and arrangements to preserve mass of drinking water within the resulting model. However, another important aspect is given by preserving demands of the original network also. As indicated in Subsection 6.2.3, it is assumed that the water supplier can provide general information about given demands with respect to each pipe element within the real network topology regarding an annual average value at least. Based on this information, a total demand regarding the whole network can be determined. Afterwards, demands can be localized by using the connection points of artificial tank elements to the network model. For this purpose, demands known for pipes considered within an artificial tank element plus demands known for nearby pipes representing paths of the final network model are related to the total demand. This means, a percentage of demands for special regions can be determined by using data given for neglected pipes collected in an artificial tank and surrounding pipes not neglected. The resulting percentage of demands is then used to define an artificial sink that simulates local water demands. In this context, an artificial sink corresponds to an external flow $Q_S(t)$ attached to an artificial tank element according to descriptions in Chapter 4, Section 4.2.2 and Section 6.2.6 of this chapter.

As the assignment of pipes and their demands to an artificial tank element corresponds to a manual approach, there is no guarantee that the resulting distribution of demands within the network model generated is completely correct. In fact, a perfect distribution can hardly be realized due to the reduced number of artificial tanks considered. This means, in the final model there will always be neglected pipes and corresponding demands considered at locations that are less appropriate because they must be assigned to artificial tanks given. For this reason, the distribution of demands resulting from the mentioned percentage assigned to each artificial tank element might be altered regarding calibration processes detailed in Chapter 7, Section 7.3.3.

6.3 Generator TWaveGen

6.3.1 General aspects

As described in previous subsections, the approach for network generation considered in the EWave project is characterized by manual procedures, particularly in order to ensure flexibility when designing a network model layout. However, having to deal with thousands of elements regarding real network topologies, such manual approaches can be quite extensive and demanding. Hence, in order to reduce working hours for aggregating a given network, support by a software tool is desirable. For this purpose, the aggregation process previously introduced has been implemented in MATLAB by realizing a network model generator for water supply systems called TWaveGen.

TWaveGen corresponds to a graphically supported program that is used to perform the steps of network aggregation visualized in Figure 6.2. So, based on given data of real topologies it allows for defining layouts, reducing pipes as well as generating artificial tanks and sinks in order to achieve a final model whose information can be used to perform simulations of hydraulic processes and conditions using programs such as TWaveSim and Anaconda that are described in Chapter 4, Sections 4.3 and 4.4. In this context, due to different requirements regarding layout definition and pipe reduction on the one hand as well as artificial tank and sink generation on the other hand, TWaveGen consists of two separate main processes. Hence, when starting TWaveGen, the user decides either layout definition and pipe reduction or artificial tank and sink generation should be considered.

TWaveGen pictures any given network topology and current states of its aggregation. Besides, it allows to show detailed information with respect to each pipe element involved such as parameters or the amount of consumers and consumptions attached. Also, it offers to include further user-defined information such as important positions that should be considered during the aggregation process. It enables to select pipes and to combine them creating superior elements until a desired number of remaining components is reached. Effects with respect to current shapes of the resulting network model are directly visualized and can be generally reversed. Thus, TWaveGen corresponds to an interactive device that ensures extensive flexibilities.

In this context, TWaveGen also allows to save current states of network aggregation anytime after a first model layout has been completely defined. This enables to pause the aggregation process and to continue the model generation later, thus, making the overall approach much more user friendly as there is no need to com-

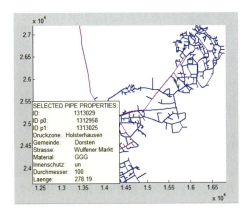

Figure 6.4. Showing network topology, selected elements and pipe properties in TWaveGen

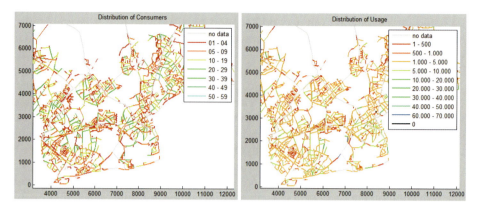

Figure 6.5. Showing the number of consumers (left) and the amout of demands (right) per pipe element in TWaveGen

plete a network model in just one single passage. This possibility of saving data with respect to current states during network aggregation also includes a straight forward procedure for realizing multiple stages of reduced models directly that could be used for comparative computations.

Handling TWavGen is simple, all activities regarding layout definition, pipe reduction as well as artificial tank and sink creation are realized and visualized within at most two graphical windows and controlled via the left or right mouse-button. The overall process generating a network model thus becomes self-explaining and accessible. Hence, working with TWaveGen demands no extensive coaching.

A major objective when implementing TWaveGen was to ease single steps of network aggregation and to reduce their effort and time-consumption by appropriate features as far as possible. In this context, especially processes that require a selection of specific pipe elements had to be simplified. Pipe selections are relevant regarding each step of the approach introduced. They are required when defining layouts, reducing pipes and creating artificial tanks. Hence, appropriate algorithms had to be realized that avoid having to select each pipe element individually. For this purpose, an automatic completion algorithm for selecting pipes when defining layouts and reducing elements as well as an algorithm to group pipes when generating tanks were implemented that allow for realizing these steps with just a minimum of mouse activities. Both these algorithms are detailed in the subsections given below.

6.3.2 Pipe selection

A relevant aspect when aggregating networks as previously described is to implement a useful procedure for pipe selection. Especially regarding the definition of an appropriate layout, a user has to deal with hundreds and thousands of pipes. Due to the manual approach in order to ensure flexibility with respect to the generation process, this means an immense effort for selecting given components. However, all elements

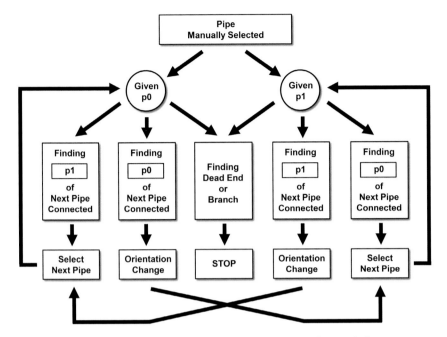

Figure 6.6. General idea of the algorithm for automatic completion

selected while defining a layout will usually be combined to a unifying element as long as they are characterized by direct connections without branch. This property can be used to speed up the procedure by automatically selecting and unifying all elements directly connected once one of its components is chosen for being included in the layout design. So, implementing an algorithm that completes a chain of directly connected elements is useful. In this context, all the pipes before and after a selected element must be automatically considered until reaching a branch or a dead end.

However, to find all elements directly connected to a single element selected is no easy task. Considering that any element within a given chain could be selected, a search algorithm for finding connected elements must run into two directions in order to select all the pipes before and all the pipes behind the element chosen. This can be effectively realized by looking at distinct identification numbers that must be provided for each initial point $p0$ and each end point $p1$ of every pipe as indicated in Subsection 6.2.3. If these identification numbers occur repeatedly for the points of different pipes, the pipes are connected. Nevertheless, although looking for all the elements given before or after a pipe selected in principal requires just to proceed into one specific direction, it is not sufficient always to consider just the identification numbers of initial or end points of upcoming elements. Indeed, both points must be taken into account with every step to the next element in order to consider the possibility of changing pipe orientations. In Figure 6.6, a scheme visualizing the general principle for realizing a corresponding search algorithm that allows for automatic completion is given.

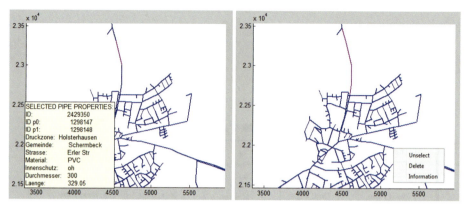

Figure 6.7. Automatic completion when selecting pipes in TWaveGen: Selected pipe (left), automatic completion (right)

Despite automatically completing given chains of pipes after selecting one of its components significantly speeds up the process of defining a layout, it might not always be a desired feature. Especially when it is required to include a special point given right within a direct connection of elements, automatic completion would cause its neglection. Hence, in order to satisfy the objective of full flexibility during the process of network aggregation, there is also the possibility in TWaveGen to deactivate the automatic completion described anytime and to continue defining a layout by selecting just single elements.

6.3.3 Tank generation

The process of generating artificial tanks that include information of all neglected pipes requires to include both data of the original network topology and data of the aggregated network model. In general, artificial tanks are realized by selecting a specified group of elements given in the original network and comparing them with information given for the aggregated model. If one of the pipes selected with respect to the original network corresponds to an element that is not part of the aggregated model, geometrical information of this element is not involved within the simulation model yet. Hence, its data must be considered when computing dimensions of the artificial tank. If one of the pipes selected with respect to the original network corresponds to an element that is part of the aggregated model, geometrical information of this element is involved within the simulation model already. So, its data must not be considered when computing dimensions of the artificial tank. Otherwise, mass of drinking water within this element would be considered twice within the simulation model and, thus, falsify hydraulic characteristics. Nevertheless, information with respect to drinking water demands are collected no matter if an element is considered in the model layout already in order to define sinks according to Subsection 6.2.7 that will finally be assigned to artificial tanks defined.

To perform this strategy, the aggregated network model must include at least information of distinct identification numbers with respect to all the original pipes involved as indicated in Subsection 6.2.3. This means, even when regarding simplified pipe elements in the final network model that result from unifying and combining pipes of the original network topology repeatedly, the knowledge of original elements within this superior pipe must be preserved. This can be realized by defining arrays during the process of layout definition and pipe reduction that collect all identification numbers. So, when selecting a group of pipes for realizing an artificial tank, the identification number of each pipe within this group must be compared with information within these arrays. Once the check was realized and a pipe of the selected group was assigned to be an element to include within the artificial tank or not, it must be ensured that this element cannot be taken into account again so that its information is not considered multiple times.

In order to simplify the process of selecting elements when realizing artificial tanks, a rectangular frame can be drawn over a group of pipes in TWaveGen. All elements within this frame are potential candidates to be included within the artificial tank and compared with information regarding the model layout generated. However, a rectangular frame is not sufficient to work precisely. Hence, further features were involved that allow to deselect pipes inside the rectangle or to select pipes outside the rectangle individually. Also, it is possible to combine a choice of artificial elements or to withdraw them.

After generating an artificial tank element, it must be assigned to one of the remaining nodes given within the model created in order to connect it with the simulation network. In general, tank realization always is given individually with respect to the model layout that was loaded for comparing with the original network topology. Thus, when realizing different stages of aggregated models there must always be another definition of artificial tanks to complete the model.

The approach introduced for realizing artificial elements is flexible. However, it also includes the risk of assigning demands and neglected pipes to improperly artificial tanks and, thus, locations. This might negatively affect hydraulic conditions. Hence, steps of calibration and validation are required that might include an adaption of artificial tank dimensions or a reordering of demands assigned. Details on this topic are given in Chapter 7, Section 7.3.3.

6.4 Conclusion and outlook

In order to satisfy strict time requirements that are given with respect to the EWave project, this chapter introduced an approach for network generation that – contrary to common methods found in literature – allows for a significant reduction of pipes in real network topologies down to a predefined number of remaining elements. In fact, the procedure introduced knows nearly no restrictions regarding the amount of reduced pipes and could be theoretically applied to any real network topology until just one single artificial tank element is left. So, leading in general to very abstract

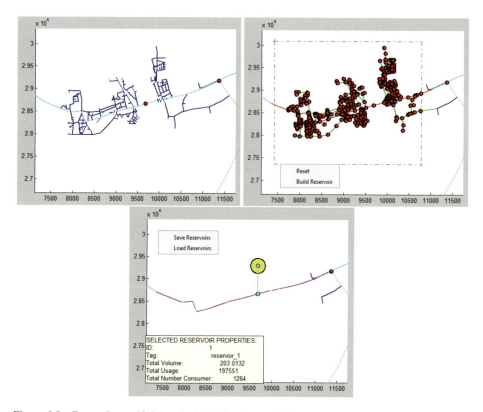

Figure 6.8. Generating artificial tanks in TWaveGen: Original network (left), group of elements selected for aggregation (right), aggregated network with artificial tank (below)

network models, the purpose of this approach is less given by preserving hydraulic conditions within every region of the resulting network model. The objective is much more to realize significant reductions in computational efforts and, thus, computing times. Hence, the focus is on creating models that satisfy hydraulic conditions just within a few number of predefined regions by appropriate calibration techniques. However, the more abstract the resulting model the harder will hydraulic conditions be satisfied, even after applying calibration strategies.

Another advantage of the given approach compared to common procedures described in literature is the extended flexibility with respect to defining layout designs of the final network model. In this context, the resulting form of the network model is entirely controlled by the user without being restricted by algorithms involved. This way, massive reductions of pipe elements can be realized while preserving the general design of the original network topologies as much as possible. However, such extensive flexibility in general increases the overall effort of realizing an appropriate model and, thus, requires to implement software tools that help to reduce the manual effort by semi-automatic strategies effectively. Some corresponding techniques have been detailed while introducing TWaveGen.

Regarding the EWave prototype region, the presented approach for aggregating networks enabled to reduce given real network topologies consisting of 15.114 pipe elements to a network model with 98 remaining pipes and 52 artificial tanks. Within this model, hydraulic conditions are preserved especially with respect to the waterworks outlets and some special sites including plants for storing water or increasing pressure by applying appropriate calibration strategies. Details on the finally resulting network model will be discussed in Chapter 7, Section 7.2. They show that the aggregation process introduced is capable to deliver very abstract network models that are, however, able to keep the general shape of the original network topology and to satisfy minimum requirements with respect to preserving hydraulic conditions.

When not being restricted to strict demands on computing times the approach for aggregating networks introduced can be used to realize more detailed models. In fact, the flexibility properties given allow to realize combinations of network regions that are drastically aggregated with network regions that are not aggregated at all, so that especially within the non-aggregated regions hydraulic conditions should be preserved without having to apply extensive calibration. Also, different aggregation levels can be realized and applied to compare results and to estimate how significant hydraulic conditions might be affected by the aggregation process.

Bibliography

[1] E. J. Anderson and K. H. Al-Jamal, Hydraulic-Network Simplification. *Journal of Water Resources Planning and Management* 121 (1995), 235–240.

[2] D. Hamberg and U. Shamir, Schematic Models for Distribution Systems Design. I: Combination Concept. *Journal of Water Resources Planning and Management* 114 (1988), 129–140.

[3] D. Hamberg and U. Shamir, Schematic Models for Distribution Systems Design. II: Continuum Approach. *Journal of Water Resources Planning and Management* 114 (1988), 141–162.

[4] D. Paluszczyszyn, P. Skworcow, and B. Ulanicki, A Tool for Practical Simplification of Water Networks Models. *Procedia Engineering* 119 (2015), 486–495.

[5] L. Perelman and A. Ostfeld, Water Distribution System Aggregation for Water Quality Analysis. *Journal of Water Resources Planning and Management* 134 (2008), 303–309.

[6] L. Perelman, M. L. Maslia, A. Ostfeld, and J. B. Sautner, Using Aggregation / Skeletonization Network Models for Water Quality Simulations in Epidemiologic Studies. *Journal of American Water Works Association* 100 (2008), 122–133.

[7] L. Perelman and A. Ostfeld, Water-Distribution Systems Simplifications Through Clustering. *Journal of Water Resources Planning and Management* 138 (2012), 218–229.

[8] A. Prei, A. J. Whittle, A. Ostfeld, and L. Perelman, Efficient Hydraulic State Estimation Technique Using Reduced Models of Urban Water Networks. *Journal of Water Resources Planning and Management* 137 (2011), 343–351.

[9] B. Ulanicki, A. Zehnpfund, and F. Martinez, Simplification of Water Distribution Network Models. In *Proc. 2nd Int. Conf. on Hydroinformatics* (A. Müller ed.), Zurich, Switzerland, 1996, 493–500.

Chapter 7
Setup of simulation model and calibration

Gerd Steinebach, David Dreistadt, Patrick Hausmann, and Tim Jax

Abstract. This chapter deals with the setup of the simulation model of pressure zone Holsterhausen. This model consists of two parts: The relevant processes within the waterworks Dorsten-Holsterhausen and the distribution network for drinking water. These parts are connected by the drinking water pumps. To consider the huge distribution network within the simulation model an abstraction is required, leading to arregated pipes and tanks. The final simulation model is calibrated. Calibration is made separately for single network elements like pumps and valves and for the aggregated network. The calibration of e.g. pumps can be done automatically, whereas the calibration of the network is largely a manual process. Finally, some typical simulation results are discussed. The achieved accuracy is appropriate for practical application and the further optimization process.

7.1 Introduction

In this chapter the structure of the simulation model for the pilot network Holsterhausen described in Chapter 2 is explained. Methods and software introduced in Chapters 3, 4, and 6 are applied for this purpose.

First, in Section 7.2.1 all relevant processes in the waterworks Dorsten-Holsterhausen are modeled according to the available network elements defined in Chapter 4. With the network generator TWaveGen (see Chapter 6) the distribution network of the pressure zone Holsterhausen is then abstracted and coupled with the waterworks model, see Section 7.2.2. Some important details of the model of the distribution network are described.

Afterwards, the complete simulation model is available and allows computations based on measured input data. For the simulation model to provide sufficiently accurate results, model calibration is required. This calibration is dealt with in Section 7.3. First, characteristic pump curves are taken into account for all drinking water pumps considered in the model. Some of the pumps are equipped with throttle valves, which are calibrated separately, see Sections 7.3.1, 7.3.2. Section 7.3.3 describes the calibration of the distribution network model. This calibration refers to the available measured values of pressure and flow rates within the pressure zone of Holsterhausen.

The different steps which lead to an improved correspondence between simulation results and measured values are described.

Finally, the operational usage of the simulation model is explained in Section 7.4 and some simulation results are presented and discussed.

7.2 Application: model Holsterhausen

The different network elements simple node, tank, pressure tank, pipe, connection, valve, control valve, pump and control pump described in Chapter 4 are available for the construction of the EWave simulation model. All processes of water production, water treatment and water distribution are described by these elements. The model construction is first made separately for the waterworks and the distribution network in the pressure zone. The coupling is carried out by the drinking water pumps, which transport the water collected in drinking water tanks from the waterworks into the pressure zone. Finally, a large monolithic network model is available for simulation. In the following these modeling processes are considered.

7.2.1 Modeling waterworks Dorsten-Holsterhausen

An overview of the main structure of the physical processes of water production and water treatment within the waterworks Dorsten-Holsterhausen is given in Section 2.2. All relevant waterworks components must be modeled with respect to their properties for network simulation and optimization. Based on the network elements defined in Chapter 4 the structure of the waterworks is built by connecting the elements to an abstracted network, see Figure 7.1. Since water quality is not considered in the model only the hydrodynamic behavior and energy consumption is of interest.

Groundwater is used as the source of water, which collects in the wells of Holsterhausen and Üfter Mark. Since measurements of flow and pressure were not sufficiently available the well galleries are not modeled in detail. They are treated as source terms delivering a given amount of water.

Water from the wells Holsterhausen is pumped into the percolator, where it is deacidified by desorbing carbonic acid dissolved in the water. The percolator is modeled by a connection with given input pressure. After this process step, the water is collected in the raw water tanks.

Water from wells Üfter Mark is collected first in an individual tank. On the way from the wells to this tank it is possible to convert a part of the energy which has been consumed for the pumping process back into electrical current by means of an energy recovery by a backward running pump. From tank Üfter Mark water is directed into the raw water tanks of the waterworks. By the control valve between these tanks the mixing rate between the water of the two well galleries is controlled. The other valves are used to model pressure loss and to allow a decommissioning of a tank.

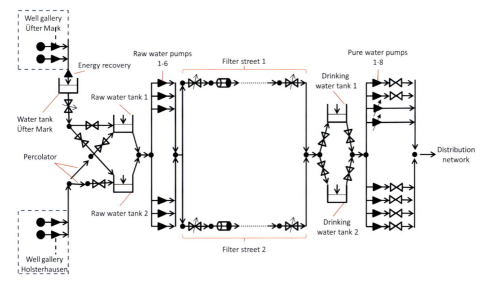

Figure 7.1. Abstracted network of waterworks Dorsten-Holsterhausen

Raw water pumps 1-6 conduct water further into two separated filter-streets combined with disinfection by ultraviolet light. The processes within these components are modeled by control-valves, pipes and connection elements. Finally, water is collected in the drinking water tanks and supplied to the distribution network by the drinking water pumps. Two of these pumps are speed-controlled to ensure desired pressure in the distribution network.

7.2.2 Modeling pressure zone Holsterhausen

In Figure 7.2 (left, blue colored) an overview of pressure-zone Holsterhausen is shown. It gets its water from the waterworks Dorsten-Holsterhausen and consists of approximately 15.000 pipes of 1.100 km total length. The zone can be subdivided into rural and urban regions. Rural regions having only few consumers can be seen by the low number of pipes and pipe-branches. The urban regions have a high population density and a higher number of pipes is installed for water supply. With respect to mathematical effort and computational cost it is not possible to consider all pipes of the network. So the complexity of the network is simplified by summarizing several single pipes elements into combinded pipes and into artificial tanks generated by TWaveGen, see Chapter 6. It has to be ensured that the main structure of the network is kept. The network has been reduced to 96 pipes and 52 artificial tanks. In Figure 7.2 (right) this abstraction is shown. At red nodes important sub structures are connected and at yellow nodes most important artifical tanks used for calibration are placed.

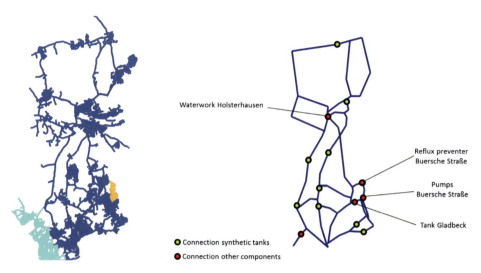

Figure 7.2. Pressure zone Holsterhausen (blue), Buersche Straße (yellow) and abstracted network (right)

Beside the abstraction of the network some further details are of interest, shown in Figure 7.3. The tank system Gladbeck consists of two tanks and four pumps. During periods of high water demand in the pressure zone Holsterhausen the pumps are in operation for increasing pressure and the filling valve is closed. During periods of low water demand filling valve is openend and the tanks are filled again.

Within pressure zone Holsterhausen there is a higher area, called Buersche Straße (Figure 7.2, left, yellow). This little area is supplied by the two pumps shown in Figure 7.3 (below).

The models of the waterworks (Figure 7.1) and the distribution network (Figure 7.2) are combined to one uniform model. The number of the various network elements are given in Table 7.1.

At the artificial tanks the water extraction according to the total water demand takes place. The fixed rates for distribution of withdrawal quantities are calculated from local annual output measurements for each street.

All pumps and some connections, representing percolators or disinfection components are supplied with a specific electric power, see Chapter 4. In this way the total electic energy consumption depending on the simulated hydrodynamic behavior can be computed.

Nodes	Tanks	Pumps	Pipes	Connections	Valves	Artificial Tanks
75	7	20	100	13	33	52

Table 7.1. Total number of elements in the simulation model

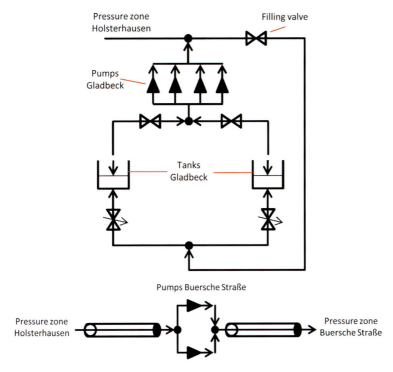

Figure 7.3. Tank system Gladbeck (above) and pressure increase system Buersche Straße (below)

7.3 Calibration

After the simulation model is set up, it must be calibrated. If sufficient measurement data is available, pump curves and valve characteristics should be calibrated first. Thereafter, the calibration of the other network elements is carried out with the aim of the best possible match between simulated and measured states.

7.3.1 Calibration of characteristic pump curves

Radial pumps are found throughout the whole water distribution network and make up the largest share in overall power consumption. To optimize the water distribution regarding energy effiency it thus is important to know about their hydraulic and electric properties. Characteristic curves describe flow rates, heads, hydraulic and electric powers as well as efficiencies and are therefore an important tool to e.g. find optimal operating points for the used pumps.

For the simulation, characteristic curves are necessary to implement the pumps into the model and represent their general behaviour, especially in interaction with each other.

Evaluated data and general characteristics. For the determination of characteristic pump curves, datasets of measured values have been provided. These datasets include values of flow rate (m^3/h), pumping head (m), electric power consumption (kW) and operating hours (h). These datasets form the basis to determine characteristic curves for pumping heads (QH-curve), electric and hydraulic powers (P_{el}, P_{hyd}-curves) and effiencies (η-curves). All curves are determined in relation to the flow rate Q, which acts as their common x-axis.

A distinctive feature of the QH-curve is it's maximum, which is always bound to the H-axis. This is due to the fact that the generated pumping head increases if the flow rate decreases and vice versa. The maximum possible head at $Q = 0$ is therefore called shut off head, since the pump no longer generates any flow. At the maximum possible flow rate (Q_{max}) the pumping head drops to $H = 0$.

Since hydraulic power is the product of flow rate and differential pressure (which is proportional to the pumping head), the P_{hyd}-curve always originates in $Q = P_{hyd} = 0$. The curve then rises to a maximum before dropping again to $P_{hyd} = 0$ at $Q = Q_{max}$.

Given that the system of electric motor, bearings, potential gearboxes and the pump itself is of course lossy, the electric drive has to generate a certain minimum power before the pump can create any hydraulic power. This causes the P_{el}-curve to cross the P-axis at a value greater than zero and is monotonously increasing.

The curve for efficiency (η) is determined as the quotient of hydraulic and electric power ($\frac{P_{hyd}}{P_{el}}$) and therefore always originates in $\eta = Q = 0$. Initially the curve rises until an optimum ratio of hydraulic and electric power is reached which results in a maximum value of efficiency. After that, the hydraulic power output starts to decrease, while the electric power input is still increasing. This causes the η-curve to fall until the pump has reached it's maximum flow rate at $H = P_{hyd} = 0$ which also leads to $\eta = 0$.

Model approach and calibration of pump curves. Based on the provided datasets, two primary pump curves are determined. First, a QH-curve based on the measured values of flow rate and pumping head. Secondly, a P_{el}-curve based on flow rate and electric power input. After that, the P_{hyd}-curve can be derived from the QH-curve and thus the η-curve can be determined as P_{hyd} divided by P_{el}. Therefore the model approach only describes the primary pump curves.

The QH-curve is modeled by a power function, see Section 4.2, equation (4.8). It ensures the maximum at $Q = 0$ and provides a good fit to the measured values [1]:

$$H(Q) = a_r Q^r + a_0 \qquad (7.1)$$

with coefficients $a_0 > 0$, $a_r < 0$ and $r > 1$.

The P_{el}-curve is modeled using a second order polynomial. It also provides a good fit to measured data:

$$P_{el}(Q) = a_2 Q^2 + a_1 Q + a_0 . \qquad (7.2)$$

As already mentioned, the P_{hyd}-curve is determined by evaluating the QH-curve and calculate the product of flow and head, although the head has to be converted to differential pressure (N/m²) beforehand:

$$P_{hyd}(Q) = Q \cdot \Delta p = Q \cdot g \cdot \rho \cdot H(Q) \tag{7.3}$$

with gravity constant g and water density ρ assumed to be constant.

Concluding, the η-curve is determined by dividing P_{hyd} by P_{el}:

$$\eta(Q) = \frac{P_{hyd}(Q)}{P_{el}(Q)} . \tag{7.4}$$

To find the correct parameter values, which means calibrating the model, the above mentioned datasets are approximated using these equations. Regarding e.g. equation (7.1), an error-function F is defined:

$$F(a_0, a_r, r) = \sum_{i=1}^{N}(H(Q_i) - H_i)^2 .$$

Here, (Q_i, H_i), $i = 1, \ldots, N$ describe measured pairs of flow rates and pumping heads and function $H(Q)$ is given by (7.1). The aim is to compute coefficients a_0, a_r, r which minimize the error function F. For this purpose least square optimization algorithms from MATLAB optimization toolbox [2] have been applied. Figure 7.4 shows the measured values and the computed functions (7.1) and (7.2). Moreover, the derived functions (7.3) and (7.4) are shown.

It is possible to use radial pumps as turbines and thereby recuperating energy of the hydraulic system by converting it back into electric energy. This happens e.g. at the Üfter Mark reservoir. To control the electric energy output the pump is installed in a bypass-pipe. Both, main-pipe and bypass-pipe can be throttled indepently, while the aim is to keep a constant differential pressure at the turbine. Thereby the turbine-power can be regulated by a flow-control.

For modeling and calibration, above mentioned formulas and methods can be utilized, too.

7.3.2 Calibration of valve coefficients

Like pumps, throttle flaps and gate valves are also widespread across the water distribution network. They open or close sections of the network and control the flow rate generated by pumps without speed control.

Therefore they are essential network elements and a proper modeling and calibration of valves is important for the overall simulation model.

For the modeling and calibration of throttle flaps or gate valves measured values of flow rate (Q in m³/s), pressure loss (Δp in bar) and degree of opening $s \in [0, 1]$ are evaluated. To allow a quick and easy comparison with e.g. measured values of pumps, pressure loss is directly converted to loss of pumping head (H in m).

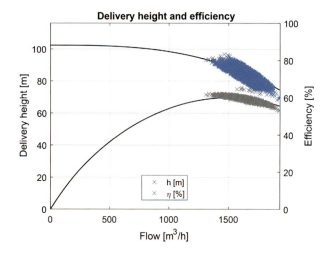

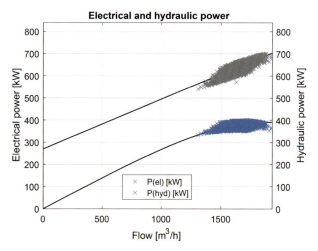

Figure 7.4. Calibration results for charateristic curves of a drinking water pump

Aim of modeling valves of the water distribution network is to reproduce the correlations between flow rate Q, degree of opening s and loss of pumping head H for the simulation.

In the model, these correlations are described by the parameter $\tilde{\zeta}$, see Section 4.2.2, equation 4.5:

$$\tilde{\zeta} = \frac{s^2 H}{Q^2} . \qquad (7.5)$$

After values of $\tilde{\zeta}$ are generated by evaluating the provided datasets using equation 7.5, it turned out that they are not constant but depend on Q and s. The following model approach with coefficients $\zeta_0, \alpha_1, \alpha_r, r_q, \beta_1, \beta_r, r_s$ led to good approximations:

$$\tilde{\zeta}_{approx}(Q,s) = \zeta_0 + \alpha_1 Q + \alpha_r Q^{r_q} + \beta_1 s + \beta_r s^{r_s} . \tag{7.6}$$

Again the least square approach has been applied for the computation of the coefficients, i.e. the function

$$F(\zeta_0, \alpha_1, \alpha_r, r_q, \beta_1, \beta_r, r_s) = \sum_{i=1}^{N} \left(\tilde{\zeta}_{approx}(Q_i, s_i) - \frac{s_i^2 H_i}{Q_i^2} \right)^2$$

was minimized. A result of this approach can be seen in Figure 7.5. Above a comparison between measured values Q_i resp. H_i and computed values acoording to equation (7.5) is shown, i.e.

$$\sqrt{\frac{s_i^2 H_i}{\tilde{\zeta}_{approx}(Q_i, s_i)}} \quad \text{resp.} \quad \frac{\tilde{\zeta}_{approx}(Q_i, s_i) Q_i^2}{s_i^2} .$$

The calibration with respect to errors in pressure loss are sufficiently small, whereas the errors with respect to flow are not completely satisfactory. Figure 7.5 below shows the dependence of $\tilde{\zeta}$ on flow Q and valve opening s. Blue points represent values $\tilde{\zeta}_{approx}(Q_i, s_i)$ and black points values $\tilde{\zeta} = \frac{s_i^2 H_i}{Q_i^2}$. Despite the not entirely satisfactory calibration of $\tilde{\zeta}$ values, a significant improvement of model results compared to constant values of $\tilde{\zeta}$ could be achieved.

7.3.3 Calibration of network model

As described in Chapter 6, aggregating network topologies is particularly based on defining a model design layout, combining pipes to aggregated elements and considering neglected pipes within artificial tanks that also consider drinking water demands by connected sinks. General purpose of this procedure is to reduce the number of elements significantly in order to safe computational effort. Because of the extensive reduction of pipes and corresponding abstract layouts resulting, preservation of hydraulic conditions can hardly be realized with respect to all regions given. Instead, the focus is on preserving hydraulic conditions with respect to flow and pressure regarding a few number of special locations where measurements are avaiable, notably outlets of the water works. Besides, reproducing water elevation levels within tanks that are really present within the original network is addressed. For this purpose, corresponding measured values provided by the water supply company must be compared with simulation results. The task then is to ensure conformity of measurements and simulation by appropriate calibration techniques that will be discussed below.

However, calibration proves to be challenging. First, because of the large number of potential calibration parameters that are given with respect to different elements involved. Second, because of the many different locations where these parameters are given.

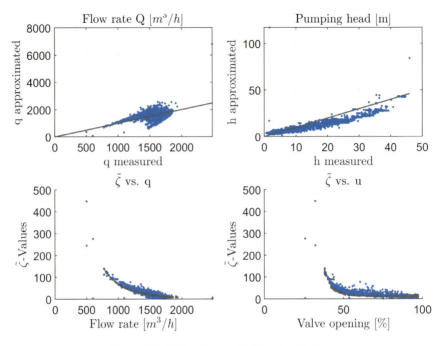

Figure 7.5. Calibration results for a throttle flap

Suitable coefficients for calibration cover e.g., friction coefficients of pipes, zeta coefficients of valves or tank dimensions that were introduced formulating corresponding model equations in Chapter 4. Therefore, contrary to the local calibration of pumps and valves a procedure for calibrating flow and pressure requires to regard the entire network model and its characteristics. In this context, especially artificial tanks reproducing volume of neglected pipes and connected sinks reproducing drinking water demands proved to be relevant for calibration. Other parameters such as friction coefficients showed just little influence on flow and pressure when trying to adjust them.

As a consequence, the approach introduced for aggregating real network topologies will lead to a first layout of neglected pipes and their demands with respect to artificial tanks that will not be able to reproduce hydraulic conditions within the network model correctly. However, it is assumed that at least pipes originally belonging to direct surroundings of artificial tank elements defined are considered appropriately. The essence of calibrating the aggregated network model thus is characterized by having to correct the distribution regarding the rest of these pipes and their associated drinking water demands by a suitable approach.

This can be realized by a manual procedure. The procedure of reordering demands assigned to artificial tanks is assumed to show influence on flow behavior primarily. Hence, an appropriate approach consisting of three steps to calibrate flow conditions will be presented first. Calibration with respect to pressure is realized by subsequent strategies.

In order to calibrate flow conditions, conformity of results given by simulating the aggregated network model and measurements given for each outlet of the waterworks into the pressure zone at different time periods should be realized. While calibration is carried out for one specific time-period first, further time intervals must be taken into account in order to realize a validation of final results.

Step1: Assigning tanks to given regions. Usually each water work outlet corresponds to a specific supply region within a pressure zone. Having detailed information of flows into these different regions by measurements, a reasonable first step for calibrating flow conditions is given by relating artificial tanks and corresponding demands to these regions. This is in general realized manually including the expertise of water works staff members that are able to evaluate which position of given artificial tanks fits to given real supply regions the most. In this context, having the possibility of including information about which pipe belongs to which region during the process of network aggregation already would be perfect. However, most often corresponding detailed information is not available. Assigning artificial tanks to given regions thus includes some uncertainty just like assigning given pipes to an artificial tank, as false decisions assigning artificial tank to supply regions are imminent.

Step 2: Reordering of percental withdrawal distribution. Based on measurement data for each water works outlet a first comparison of percentaged drinking water withdrawals can be realized. In general flow and withdrawal are directly connected, meaning that there are higher flow rates to regions characterized by higher drinking water withdrawals. Regarding artificial tanks, withdrawals are given by drinking water demands. Hence, by adapting the percentaged distribution of drinking water demands assigned to artificial tanks the overall flow behavior within a network model can be manipulated and thus improved. In this context, the percentaged distribution of demands belonging to a subset of artificial tanks that were assigned to a specific supply region by the previous step should reflect the percental withdrawal known for given water work outlets by measurements, as every outlet usually belongs to one of these specific supply regions.

An appropriate first step of calibration is given by adapting the drinking water demands assigned to artificial tank elements by connected sinks such that they correspond to total values resulting from measured data for the water work outlets. This can be applied by manual approach in different ways, e.g. by manually decreasing or increasing the percentaged demand for each artificial tank element assigned to a specific region. In this context, each step of correction must ensure that the total demand known will not be decreased or increased. Also, it must be ensured that negative values are avoided while trying to preserve the total demand, i.e., sinks assigned to artificial tanks are not allowed to work as additional wells that produce additional amounts of water into the system.

By comparing the percentaged distribution of drinking water withdrawals, a first correction of flow behavior within the system is in general realized.

Step 3: Further strategies for calibrating flow rates. The process of calibration by comparing measurement values with simulation results is repeated until an adequate sufficient conformity is reached. For this conformity no sharp values exist. It is up to the user to decide, when a good agreement is achieved. However, in some cases adjusting demands assigned to artificial tanks might not be productive enough. In these cases, further calibration techniques must be applied. Possibilities are given by changing coefficients of pipes, valves and artificial tank elements, e.g. friction coefficients. However, influence of adjusting corresponding values proved to be less significant.

Alternatively, changes regarding the characteristic structure of a given network could be taken into account. Trying to preserve flow conditions at least with respect to given water work outlets, it can be helpful to change paths within the resulting network model. In general, pipes within the final model layout are responsible for connecting artificial tank elements and thus regions given. However, during the aggregation process, paths could have been constructed that lead to inappropriate connections. By manual approach, tests can be realized in which slightly adapted network structures are analyzed. These might be characterized by neglecting some of the connections given and might lead to far better results.

Calibrating pressure. After calibrating flow, calibrating pressure can be realized by appropriate strategies. These usually concentrate on adjusting pressure targets of speed-controlled pumps and $\tilde{\zeta}$-coefficients with respect to artificial tank elements that are known to have influence on pressure distributions. In general, calibrating pressure is characterized by a similar approach as considered for calibrating flow, i.e. simulation results with respect to specific positions within the network model are compared to corresponding measured values provided by the water supply company. Due to extended abstraction of network models, it might be appropriate to concentrate calibration of pressure distribution to a selection of special locations as realized for calibrating flow conditions. Such locations could be outlets of waterworks and/or known plants for changing pressure conditions or any other locations where meausrements are collected.

Contrary to calibrating flow the process of calibrating pressure proves to be less complex as the number of elements having a significant impact on pressure distribution is quite small. Indeed, pressure targets of speed-controlled pumps prove to be most relevant, while adjusting $\tilde{\zeta}$-coefficients of given artificial tank elements might affect pressure within specific regions just slightly.

Besides regarding flow and pressure with respect to specific regions, notably waterwork outlets, the process of calibration and validation can be theoretically applied to any other position within a given network as long as measurement data for corresponding points is available. These positions must be known a priori in order to consider them during the process of network aggregation so that they are involved within the layout of resulting network simulation models. However, regarding an extended level of network aggregation that corresponds to a significant abstraction, chances of successfully calibrating flow and pressure within these points are usually marginal.

7 Setup of simulation model and calibration   141

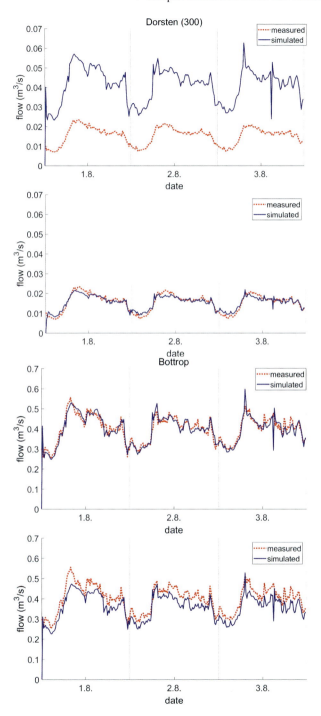

Figure 7.6. Comparison of flow results before and after calibration: First and second picture: Dorsten before and after calibration, third and fourth picture: Bottrop before and after calibration

Validation. Based on determined calibration parameters a validation of given results can be performed by comparing simulation results with measured values regarding further time periods. In general, validation requires just a little number of additional steps for adjusting parameters of given artificial tanks once the previous calibration process led to satisfying results already.

Application. Calibration strategies introduced were applied to pressure zone Holsterhausen regarding the network model presented in Section 7.2. Special focus was on preserving flow conditions with respect to the five waterwork outlets that lead to supply regions called Dorsten (pipe outlets of diameters 300 and 500 mm), Bottrop, Gladbeck and Wulfen. Calibration of flow was primarily realized by reordering demands assinged to artificial tanks as previously described. Pressure was calibrated mainly by adjusting the pressure targets of the speed-controlled pumps in order to meet the measured pressure behind drinking-water pumps given at the end of the water work. Exemplary results are shown in Figure 7.6 regarding flow conditions and Figure 7.7 regarding pressure conditions. As can bee seen in Figure 7.6, calibration leads to a big improvement of flow results concerning Dorsten (300). On the other hand, this improvement worsen a little the results concerning Bottrop which were very good already before the calibration process. Figure 7.7 shows, that simulation

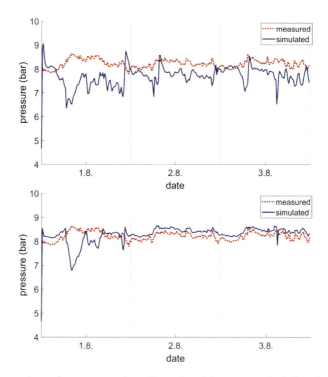

Figure 7.7. Comparison of pressure results at the outlet of the waterworks before (upper) and after (lower) calibration

reasults for pressure at the outlet of the waterworks could be imporved, too. Nevertheless, some differences between measured and simulated values remain. It should be noted, that the measurements are are also subject to some measuring tolerances.

For calibration a time period of three days in August 2016 was used to compare simulation results with measurement data given. Validation was given with respect to a time period of three days in November 2014. Overall results turned out to be promising.

7.4 Model operation and simulation results

After successful calibration, the simulation program TWaveSim (see Chapter 4) can be applied operationally. Usually, simulation is applied first to compute consistent states of the entire network that match avaiable measurements. This simulation covers the last one to two days until the current time t_0.

Thereafter, the optimization modules are applied to calculate pump schedules and valve controls for the next $n = 12$ or $n = 24$ hours. These results are passed back to the simulator and a detailed simulation of the hydrodynamic states and energy demand for the period $[t_0, t_0 + n \cdot 3600s]$ takes place.

The next Section will first describe the program handling and required input data. Finally, some results are shown exemplary in Section 7.4.2.

7.4.1 Input data and program handling

In order to start simulation, input data must be provided. This input data consists of pump schedules, settings for valve openings, total water demand of pressure zone Holsterhausen and water deliveries by well galleries Üfter Mark and Holsterhausen. Usually, all data is time-dependent. In Table 7.2 an example for pump schedules is given.

Simulator TWaveSim can be startet in two modes: Simulation of past time period and forecast mode. In the first mode all input data is obtained from measurements.

Year	Month	Day	Hour	Minute	Second	Pump1	Pump2
2016	6	19	14	31	0	0	1
2016	6	19	15	1	42	0	1
2016	6	19	15	2	42	1	1
2016	6	19	15	3	10	1	1
2016	6	19	15	4	10	1	0
2016	6	19	18	55	49	1	0
2016	6	19	18	56	32	1	0
2016	6	19	18	56	49	0	1

Table 7.2. Schedules for two pumps

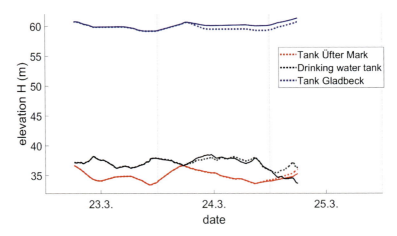

Figure 7.8. Measured (dotted) and simulated (solid) water elevations at different tanks

Moreover, input data should be completed by measurements of water levels in the tanks. Through this additional data it is possible to adapt simulated water levels to the measured ones by introducing artificial external flow in each tank, see Chapter 4, Equation (4.18). This is an important issue, especially when starting the simulator the first time and to prevent the divergence of measured and simulated states [3].

Before the simulator can operate in forecast mode a short past time period, i.e. one or two hours, prior to that forecast time is simulated. Initial conditions for that run are taken from the previous run in simulation mode. During the forecast mode external flows are switched off and input data for water levels in tanks is not required. The water demand is provided by the demand forecast programm TWaveProg, see Chapter 3. From this overall demand is derived which water quantities are to be deliverd by the well galleries. Pump schedules and valve openings are provided by the optimization algorithms (see Chapter 5), or alternatively manually.

7.4.2 Discussion of simulation results

In the forecast mode, a prognosis of the water demand by TWaveProg and pump schedules and valve controls by optimization are taken as input data. Since the forecasted demand is not exact and the waterworks is never operated exactly according to the proposed control of the optimization, a comparison between simulated and measured results is hardly possible. Therefore this comparison is restricted to the simulation mode, based on measured input data.

As a typical example, a simulation over two days, starting at 23.03.2017 06:00, is shown. During the first 24 hours artificial flow terms at tanks, where measured values are available were switched on. Figure 7.8 shows a good agreement between measured and simulated water elevations during that time at different tanks, compare Figures 7.1 and 7.3. For the next 24 hours, these artificial flow terms were shut off and small differences between measured and simulated values arise.

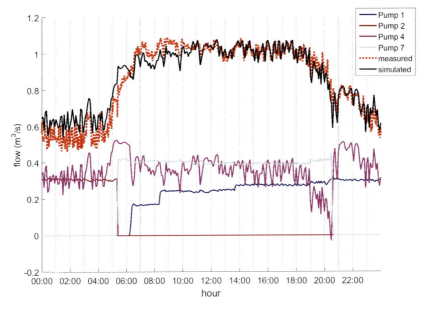

Figure 7.9. Simulated and measured flow at outlet of waterworks and throughput of different drinking water pumps

For a better overview, in Figure 7.9 flow results are shown only for the 24th of March. One can see the flow through different drinking water pumps, compare Figure 7.1. It is clearly visible that pump 4 is a control pump. The sum of all pump flows is displayed as the black line. This is compared to measured data, shown as red dotted line. It should be noted, that no information on the distribution of the actual water demand over the whole network is available. This distribution was estimated from annual data and finally corrected by the calibration, see Section 7.3.3. Despite this lack of information, measured and simulated flow values at the outlet of the waterworks show a pretty good match.

7.5 Summary

Within this Section the structure of the simulation model for pressure zone Holsterhausen was discussed. All hydraulic processes within the waterworks could be modeled using the introduced network elements. In order to achieve approriate computing times and to get a reasonable simulation model, large abstractions of the dricking water distribution network were necessary.

Calibration of the final model was divided into two stages. First, all pumps and relevant valves were considered. By least squares minimization, appropriate characteristic curves could be derived from measurements. Second, in Section 7.3.3 simplified strategies for calibrating flow and pressure within a network model were considered. Because of the large number of potential parameters for adjusting flow and pressure a manual approach was considered that is based on comparing simulation results with measurement data. Due to immense abstraction levels usually given for corresponding network models, the process of calibration and validation was especially concentrated on a few number of special locations, notably the outlets of given water works. In this context, especially flow conditions could be calibrated by re-ordering the percentaged distribution of drinking water demand assigned to artificial tank elements constructed.

The manual approach introduced led to satisfying results. But it can be a costly process, especially when having to deal with a large number of artificial tank elements. However, automating steps of this approach for future applications seems to be theoretically possible. For this purpose, an iterative procedure could be implemented that first compares percental distributions of drinking water withdrawals measured at water work outlets with drinking water demands of sinks assigned to artificial tanks. Then, this procedure could reorder the distributions of all tanks belonging to a special supply region until their amounts equal the withdrawals of corresponding water work outlets. Afterwards, further reordering could be applied by comparing results for measured and simulated flow conditions, i.e. regions with less flow rates simulated than measured require higher demand and vice versa. Nevertheless, implementing appropriate strategies for reordering demands of artificial tanks might be challenging.

Finally, in Section 7.4 some typical simulation results were discussed. The achieved accordance to measured data seems sufficient for practial applications and for the otimization process based on this simulation method.

Bibliography

[1] C. Hähnlein, *Numerische Modellierung zur Betriebsoptimierung von Wasserverteilnetzen*, PhD thesis. Technische Universität Darmstadt, 2008.

[2] *Mathworks*. https://de.mathworks.com/products/optimization.html. 2017.

[3] G. Steinebach and K. Wilke, Flood forecasting and warning on the River Rhine, *Water and Environmental Management* 14(1) 2000, 39–44.

Chapter 8

Field data, automation, instrumentation and communication

Constantin Blanck, Stefan Fischer, and Michael Plath

Abstract. For the analysis, existing data were first digitized and made available to the project partners for model construction. The system data in the waterworks as well as the data of the pipe network are to be named here. Measurements were retrofitted at various points in the waterworks and in the pipe network. The recorded operating data was then made available to project partners for the calibration of the models and for test runs of the EWave system.

8.1 Data provision, network analysis and acquisition of additionally required measurement and control technology

For an analysis of the operating states in the network, the provision of various data is necessary. In addition to flow rates, pressures and energy requirements, which are already being measured, a large amount of static data also had to be made available.

Therefore, the required data can be divided into two categories. On the one hand, there are the higher-level general data, and on the other hand, the measured operating data. The general data can be regarded as fixed, unchangeable technical parameters. This includes the geodetic heights and the geometric dimensions of the individual plant components. This data is known to most water supply companies. However, it is problematic that not all water supply companies have digital access to this data. In these cases, it takes a lot of time to look through plans and collect and summarize the data. Many geodetic heights, e.g. of retrofitted pressure measurements, must then still be measured. The correct interpretation of the pressure measurement data is only possible in conjunction with the geodetic height. Below is a list of all operating data:

- wells
  - number
  - height of the groundwater level/standing tube level height
  - vross-section and well depth
  - measured filling levels

- pumps/blowers and energy recovery
  - number
  - flow rate
  - discharge head
  - energy demand/gain
  - efficiency
  - characteristic (measurement of pressure and throughput)
  - mounting height or installation height
  - pump schedules
- multi-layer filters and UV disinfection
  - number
  - dimensions: area/cross-sectional area, height
  - ground level
  - flow rate
  - pressure losses
- trickling water, raw water chambers, water and drinking water tanks
  - number
  - area/cross-sectional area
  - ground level
  - minimum/maximum filling level
  - water levels/fill levels of tanks
- geo. heights, installations, pressure measuring points etc.
- operating hours of aggregates
- network data (pressure and volumetric flow)
- limits (levels, proper operation of pumps, etc.)
- on-site regulation
- automation

The measured operating data described above were made available to the project partners for different periods of time and in varying degrees of detail. A complete list of the transmitted data is not provided; only two examples are given for clarification. In the middle of 2015, the Bonn-Rhine-Sieg University of Applied Sciences was provided with delivery data for the preparation of the water demand forecast. In August 2016, minute values over the period of two months were sent to the project partners to validate the model of 41 data points at the Dorsten-Holsterhausen waterworks. This corresponds to approx. 3.6 million values, which were further processed.

In addition to the data of the waterworks described above, data of the distribution network were also required. In the beginning, the data of the pipelines of the pipe network connected to the Dorsten-Holsterhausen waterworks were handed over to the partners. This consists of 29,666 data sets from a geo-information system with coordinates of start and end point of the line, diameter, information on material, road, etc. Further data was later exchanged. Right at the start of the project, pressure and volumetric flow measurements were installed in the distribution network. The data recorded here were used to calibrate the model. In the further course of the project, the measurement data was transmitted to the project partners and processed.

8.2 Electricity price forecast

At RWW, forecasts were made on the one hand regarding the history of electricity prices and on the other hand regarding the expected in-house electricity generation by the Kahlenberg power plant. Short-term electricity price forecasts can be purchased on the market; there are various providers for this. Most of these companies are active in the field of "virtual power plants". Since RWW currently has a fixed electricity price for the year, the extension of the EWave decision support system by a variable electricity price over the course of the day would not have been expedient. Switching operations would then possibly even have had negative effects. Furthermore, an attempt was made to predict the generation capacity of the Kahlenberg power plant with appropriate accuracy. Among other things, water levels and flow rates of the Ruhr above the power plant were used for this purpose. A connection, possibly also with a time delay, could not be established. This can be explained by the barrages of the Ruhr and the influence of the underwater level at the power plant. In addition to the volumetric flow rate, the drop height is decisive for energy generation in the power plant. If the water cannot drain off and accumulates in the underwater, the drop height and thus also the generation capacity is reduced. One possible reason, for example, could be a flood in the Rhine which causes a correspondingly higher water level in the underwater of the power plant It seems almost impossible to include all influencing factors in the forecast.

Chapter 9
New ICT architecture

Tim Schenk, Moritz Allmaras, Andreas Pirsing, and Annelie Sohr

Abstract. In this chapter, we discuss which indicators and views of water supply systems are needed to address all relevant user roles targeted by a decision support system. We begin with an overview of the indicators that are used to estimate and compare the efficiency of water supply systems. The specific energy consumption is only of limited usefulness in this context, hence hydraulic and electric efficiency indicators are defined that allow the formulation of an overall efficiency of a plant. Then, the relevant user roles and related quality attributes for a decision support system are investigated. The chapter concludes with a detailed discussion on how the user interface for the EWave system has been designed and implemented in order to meet these system qualities.

9.1 Requirements

Considering mainly the functional requirements, the EWave project aims at the development of a software application that foremost provides the operators of a waterworks and distribution network with recommendations how to increase the plant operation efficiency. These are cyclic recommendations of setpoint values of controllable components to enable a more resource efficient control of the complete water supply cycle. The last chapters gave an insight what different types of data and calculations must be considered when a deployed software program shall be capable to really produce reasonable applicable results and not only show the theoretical potential of mathematical algorithms like simulations or optimizations. Different kinds of prognosis data must be taken into account, as well as measurements from field sensors, regulatory and technical boundary conditions, and presettings from the operators. Each data source provides data in different formats regarding data structure, data units, resolutions and further attributes. Some of the data processing modules like simulations and optimizations operate typically model-based, others are more unstructured like sensor measurements.

To realize this complex interplay of data and evaluations in an online environment, the common approach is to implement proprietary software. In doing so, the architecture is designed in a pure solution-oriented way, which ensures the application to perform its required functional task(s) for a single specific water supply plant and network – all evaluations are integrated, and their interplay is implemented in a proprietary manner. Issues with this approach are that the software is not designed for any change and it is not easily applicable to other water supplier sites and networks.

In the era of *Industry 4.0* and the *Digital Twin* and the impacts of these approaches on simulation [5], this obviously cannot be the approach to follow. In the recent decades software techniques that allow much more modular and flexible architectures have been developed. There are programming paradigms like object-oriented programming, there are new communication technologies across networks like TCP/IP protocols, HTTP protocols and socket techniques, and regarding system development, there are model-based approaches to model and evaluate complex physical systems. Thus, the design of the EWave architecture should take non-functional requirements regarding modularity and flexibility into account:

- It should be flexible in attaching and detaching software modules each implementing a specific evaluation or calculation. Especially this requirement emerges implicitly in the EWave project as the different involved teams developed their calculation modules on their own. So, there was the need for a plug & play concept to assemble all the single software pieces.

- It should be capable to use remote calculation modules, so the implementation of the modules can be realized in any kind of environment, ranging from the usage of different programming languages up to running on different operating systems or even on a different server. This requirement is also already present in the EWave project setup, as the optimization algorithms are running in a Linux environment, whereas the simulations are taking place in a Matlab runtime [6] which runs on Windows.

- The concept of a single source of truth [12] shall be considered and adapted to the EWave data handling throughout and beyond an EWave cycle. This implies that at least the topology and all physical parameters of the water supply system are present in one model and all the calculation modules rely on that model. It is ensured by design that if changes to the model occur by any participant all other participants immediately also will use that changed model.

Respecting these non-functional requirements, the EWave architecture can serve as a basis for the realization of various such resource-optimization projects with respect to different topologies of water treatment plants and networks and different evaluation participants, e.g. different prognosis modules or different measurement providers. From a high-level perspective, there are still several ways to design such an architecture. Two of the probably most intuitive are the central oriented and the – in *Industry 4.0* most propagated – decentralized approach. The central approach still has one or more central units that manage communication and distribution of data between all participants. In the decentralized approach, implemented often by a (micro-)service approach [8], all participants are fairly equal and take part in a global system. Each participant offers specific services and consumes data from other services to perform his own evaluation, like in [3]. This approach is mainly used when there is a huge number of continuously changing services that are not known a priori, each providing simple data sets and working very independently from each other. In EWave the number of participating software modules is fairly small, the generated data by single module executions can be quite big and the sequence of calculations has to take place in a fixed predefined order.

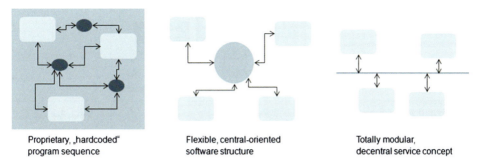

Figure 9.1. Different software structures

So, the paradigm for the design of the EWave architecture was to be modular and flexible in terms of participants and data, instead of a proprietary software solution, but still, establish a central management unit to coordinate all the communication and data exchange [see Figure 9.1].

Reflecting the third non-functional requirement, the *single model of truth*, a further aspect must be considered. The target deployment of EWave is a software system that is running online in parallel to the real plant and thereby acts on that single EWave model. Although the structure of that model is specifically designed for the utilization in such an operational support system, it is similar to models created and used in the engineering phase and utilizes a lot of information generated there. This adds a lifecycle perspective for the *single model of truth* that spans over at least the phases of engineering, commissioning and operation. This lifecycle consideration has been outlined in detail in [11] and finds a realized utilization in the concept of the EWave system architecture.

9.2 General architecture and model-based approach

9.2.1 Overview

A single entire EWave evaluation processes all participants in the predefined sequence. For each participant, this means the simple procedure of gathering its necessary input data, triggering the execution of that participant and processing its results is performed. As this whole sequence is called periodically (see Section 9.3.8) a single sequence evaluation is also called a single cycle. The overall architecture of EWave (Figure 9.2) is designed to facilitate the realization of such a cyclic sequence evaluation and therefore covers four main parts:

- The user interface. The main purpose is to show the results that are relevant to the operator. There may be also the possibility for the user (maybe not the operator but another user role) to execute sporadic actions, e.g. load historic results.

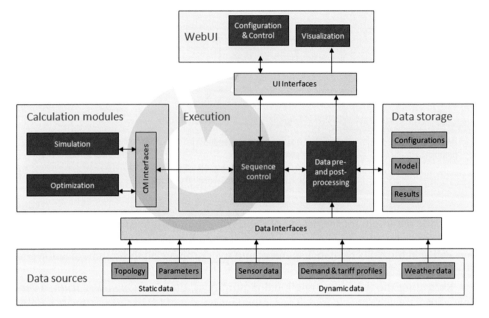

Figure 9.2. Coarse architectural overview

- An interface for calculation modules. The modules themselves use mathematical algorithms, are mainly model-based (e.g. simulation, optimization) and evaluate different necessary intermediate or final results of the overall EWave goal of finding the most resource efficient setpoints for the whole optimization horizon.
- A data management that enables the load and save of different types of data from different resources, as well as their processing and preparation for the utilization by the calculation modules.
- And finally, the central sequence control, which – based on a configuration – manages the correct sequence of the EWave evaluation by calling of the different participants (calculation modules and data sources) with the appropriate sets of data.

Up to this point, the EWave architecture could be the basis for any operational decision support system, whose evaluations comprise multiple different calculations in a defined sequence and take various data sources into account. Apart from this first main aspect, EWave is based on the second main aspect of utilizing a model-based approach and thus implicitly supporting model-based evaluations like simulations and optimizations.

Simulations and optimizations of complex physical systems that consist of multiple instances of a much smaller number of different types of components are today typically implemented in an object-oriented way [2]. Commercial simulation tools like Matlab Simulink [6] or Amesim [1] and the Modelica language [7, 10] are exam-

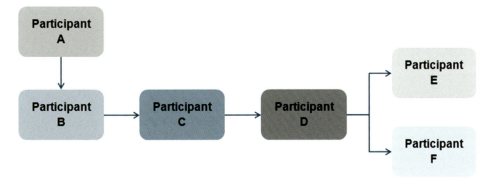

Figure 9.3. Process flow

ples that use such a modeling approach. The general idea of this approach is to define basic libraries of component types, generate a specific system model by instantiating individual elements and connecting them and implement the mathematical solvers to be independent of the number of instances and connections. This concept is also applied to the EWave data management, the sequence control and the interplay of all participants.

9.2.2 Process and data flow

A complex decision support system like EWave must evaluate a sequence of different process steps for each assistance functionality it offers, like the calculation of optimized setpoints. Each of the single process steps can be a complex model-based calculation, a pure data import or any analytical function. The sequence follows a predefined schedule of all the participants, a simple example could look like in Figure 9.3.

Each of the participants requires a specific set of input data and computes it's also specific set of result data (output). The set of input data is typically not a 1:1 equivalent of the output of the predecessor in the sequence but a certain mix of the results of all predecessors. Thus, the data flow can be very different from the process flow as depicted in Figure 9.4.

There are commercial tools that support this kind of process and data workflows for algorithmic and simulation-based workflows, like HEEDS [4] or Model-Center [9]. But they lack a general approach for the data model that is used and exchanged as they are not focused on component-oriented systems but arbitrary calculations and mainly 3D simulations. The EWave system restricts the exchanged data to be component-oriented and thus enables the ability to be completely generic in the data model. Through a whole single EWave evaluation (and as will be shown later even beyond that), the process flow is based on a single data model. Every participant retrieves the portion of data it requires as input set and contributes to its result set to the enrichment of the overall data model. Finally, at the end of the se-

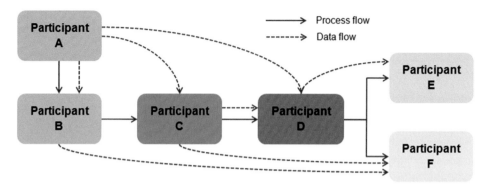

Figure 9.4. Process and data flow

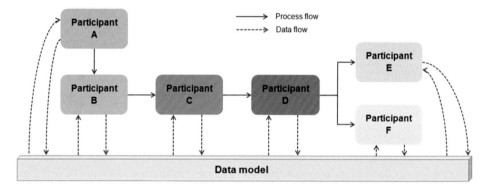

Figure 9.5. Single model of truth approach

quence, all results are present in the data model. Thus, the data flow does not take place between the different participants but from and to the overall data model (Figure 9.5).

9.2.3 Structure of the data model

To be able to act on the one postulated model, there has to be an agreement on how the model is structured and what it contains. The structure of that model is derived from the different kind of data that component-oriented evaluations require and produce.

In this data model, a plant (or more general, a system) is described as a set of components and connections. Each component has a type from a list specified in a library definition and a unique ID. It is not allowed to have two components with the same ID in the same plant. The library definition will be explained later in this chapter.

The connections describe how components are connected to build a whole system or network. A connection always links a port of one component with a port of another component, i.e. the ports are the connection points of the component. It is specified in

Figure 9.6. Component-oriented modeling

the library definition which ports a component has, of which type they are, if they have a direction and if they are single- or multi-valued. Port types specify how a connection must be interpreted, e.g. as a hydraulic connection or as a signal connection. For signal connections, the direction is crucial, while for hydraulic connections there is no distinguished direction, the direction of the flow results from the pressure conditions. The type of connected ports have to coincide, the directions have to match. A single-valued port allows only one connection linked with this port, a multi-valued port also allows more than one connection.

A special kind of component-oriented model often used in simulations is a graph model with nodes and edges. This model contains a set of nodes and a set of directed edges that connect the nodes. It can be seen as special case of an EWave component model as described above.

- Any node type may be represented by a component type with one hydraulic, multi-valued, non-directional port.
- Any edge type may be represented by a component type with two hydraulic, single-valued, non-directional ports.
- Connections always link components representing nodes with components representing edges.

Since the EWave hydraulic simulation and optimization modules both use a nodes & edges approach, the EWave data structure allows for both models: the nodes & edges model and the components & connections model. Transformations from one structure to the other are implemented. In the following, we concentrate on the components & connections model.

So far, the data structure was illustrated by hydraulic components connected to build a hydraulic network. But also, the energy supplier and even more abstract, intangible assets like an energy consumption and a boundary condition for the optimization are provided as components. These components are not connected in a physical manner but in some way assigned to other components. For this reason, the data structure allows assigning components to other components and thus introduces some kind of component hierarchy.

Components, connections and assigned components are the elements that completely define the network topology and interrelations. To specify the details of the component characteristics and behavior, the structure introduces parameters, states and profiles as elements of the components:

- Parameters are used to specify the static characteristics like geometrical data of pipes and tanks, technical data of pumps, material data, and further physical data. Parameters can have different formats (integer, real, string, boolean) and can be a scalar, a vector or a matrix.

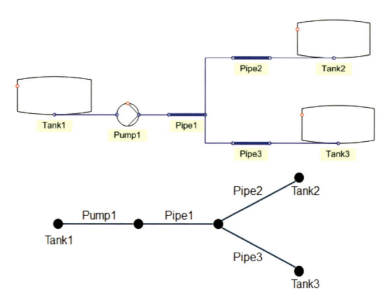

Figure 9.7. 7-components-model is transformed to nodes & edges model with 5 nodes and 4 edges

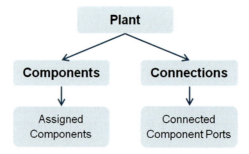

Figure 9.8. Data model for plant topology

- States are used to specify temporal characteristics like the initial filling level of a tank before simulation or the final filling level after simulation. They differ from parameters by additionally having a time stamp.
- Profiles are used to specify a behavior over time, for example dynamic loads as input for a simulation or a pump schedule as output of the optimization. These time series can be defined by pairs of time and value or as series of values for equidistant points in time. As parameters and states, they can have different formats and dimensions.

The set of parameters, states and profiles a component can have is type-specific and thus again specified in the library definition.

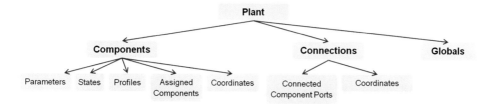

Figure 9.9. Data model

Any component and connection independent from its type can also contain information relevant for its visualization. This includes both geospatial coordinates which allow for a map visualization and drawing sheet coordinates for a schematic flow sheet visualization.

Finally, the data structure also provides global parameters, states and profiles which are not component-related but of relevance for the whole system. They may be referred by any component.

The EWave data model is designed to be a common concept for all kinds of model-based decision support systems. To make it applicable to a domain or application-specific decision support system, a definition of a domain library is necessary. This library definition concretizes the EWave data model. It declares

- on the one hand, the specification of all available elements like
    - the set of possible component types,
    - for each component type, the node-and-edge specification, if it is to be treated as node, as edge or still as component, when the node & edge representation of a specific plant model is built
    - for each component type, the set of ports (and their specifications)
    - for each component type, the set of parameters, states and profiles (and their specifications)
- and on the other hand, for each specific call to a calculation module (calculation modules can offer different possible types of executions), the set of component parameters, states and profiles, which are required as input to the module and those which are delivered as output. Conditional inputs and outputs that depend on the value of certain parameters are also possible.

That means, the library does not only define all available components and their parameters, states and profiles but also the data flow through the execution sequence. Figure 9.10 shows the high level component definition without the specific attributes for ports and parameters, states, and profiles of the EWave Valve.

Using the components and connections from the EWave library definition, the EWave pilot plant model for the selected waterworks and network distribution zone is then created by instantiating the library elements and parameterizing them.

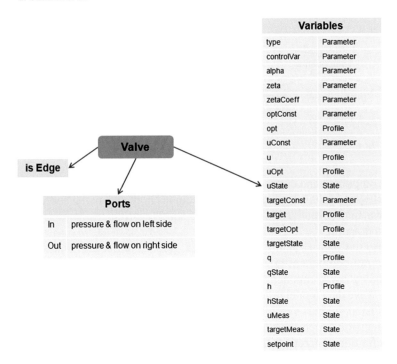

Figure 9.10. Example library definition of EWave valve

Figure 9.11 displays a small extraction of the specification of inputs and outputs of the simulation state calculation. As these inputs and outputs refer by design to the library definition, the EWave sequence control can therefore be implemented in a generic way, retrieving the data flows for the execution sequence from the library definition and applying it to the EWave pilot plant model.

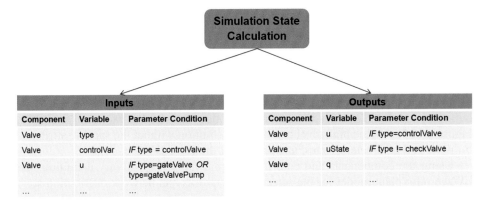

Figure 9.11. Specified inputs and outputs of the simulation state calculation

9.2.4 The calculation module interface

As mentioned before, one cycle in the EWave evaluations consists of a sequence of calculation module calls. In this sequence, a module can be called in different cases. For example, the simulation state calculation from above states a call to the simulation module for the case of state estimation. From the EWave perspective a certain calculation case of a calculation module defines itself not necessarily by the use of different algorithms or course of action inside the calculation module, but more by the different external data flows and thus the input and output specifications. Some calculation modules, like for example the EWave optimizer, only have one case, e.g. the optimization of the setpoints. Other calculation modules can have more, for example the EWave simulator can

- either simulate from the past, using historic measurements data, to calculate the current state of the network for the current time or
- be triggered to simulate the future behavior of the network over a certain time horizon.

Both cases use the same simulation module, but the required types of input data and produced results are quite different. Thus in EWave the simulator has two calculation cases with different input and output specifications.

The structure of the interface of the calculation modules is derived from the required functions and interactions with the EWave execution system which a module has to perform:

- The module has to be identifiable by the EWave execution system, therefore it has to carry and provide its unique identifier.

- The module should be able to report its internal status, consisting of a status value of defined format, possible messages and further information, e.g. the completion percentage of its calculation. It shall be possible that both the module can provide the status information itself and the execution system can ask the module for its status.

- Typically, complex calculation modules, like simulators or optimizers, require a one-time initialization of the internal structure. So, it should be possible to trigger this once before the sequence flow of calculations is started. Besides the pilot plant model, this call can require the values of internally used calculation module parameters, like solver attributes, time processing properties or error strategies.

- The execution call to a calculation module provides the module with a plant model containing all input values and a value-empty plant model with all the output variables. As for the initialization, specific values of calculation module parameters may be also necessary.

- As stated at the beginning of this chapter, the EWave calculation modules can be of various different types and therefore, be usable only in specific runtime environments, meaning different operating systems and different dependency settings. It also should be possible to build up a distributed execution system,

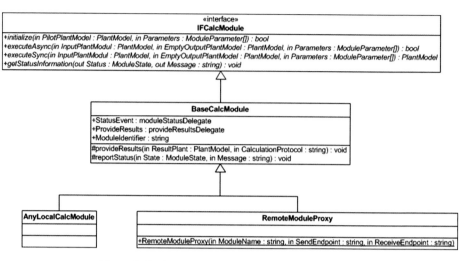

Figure 9.12. Calculation module interface specification

where the modules are not located on the same hardware as the execution system. Thus it has to be possible that calculation modules can run remotely.

- Remote calculation modules and possibly long evaluation times of single modules, while other modules could already be executed, require the possibility to call calculation modules asynchronously. This also implies that there is a way for the modules to provide their results at the end of their calculation.

The resulting specification for the interface and the basic calculation module is depicted in Figure 9.12.

Local calculation modules can be derived directly from that basic module.

Additionally, an implementation has been realized that allows remote invocation across a TCP/IP network connection. Hence, the EWave system allows individual calculation modules to be deployed across different machines and which communicate to the sequence control component via networking. This approach affords several benefits:

- Different modules can run on different platforms with the only shared dependency being the common contract given by the calculation module interface.
- The boundaries of modules are clearly defined, dependencies are isolated and cannot conflict with each other.
- Modules can be developed, deployed, and tested independently allowing different teams to work on different modules with minimal overlap.
- During operation, modules can be scaled independently depending on the workload requested. This benefit has not been directly exploited by the EWave system as its focus is the single pilot network of the project, but could be highly beneficial when operating a decision support system that serves multiple water supply networks.

- Likewise, the computational burden of individual calculation modules can be distributed across multiple machines.

The downsides to this approach is the increased overall complexity of such a distributed system, the more complicated orchestration during operation and the relatively slow performance of network communication when compared to shared memory architectures. However, in the EWave project the benefits outlined above have been found to outweigh these downsides. The distributed system approach also is a good fit with the decision to implement a web-based user interface (see Chapter 10) as such user interfaces naturally use TCP/IP-based protocols for communication between client (web browser) and server (EWave web server).

For such remote modules, an implementation of a remote proxy that provides already the remote running module has already been derived from the basic module.

For transport and message framing, the ZeroMQ library [13] has been utilized. ZeroMQ implements a message-based request-reply pattern based on a lightweight wire protocol and is available as open source software for various platforms. The exchanged messages consist of xml strings that follow a defined xsd structure (see Figure 9.13):

- The main container is the calculation module message structure that can contain either a status message or a control message.
- A status message contains information about the calculation module status.
- A control message comprises the type of action that is requested, the relevant plant data in the data model structure and the actual configuration settings.

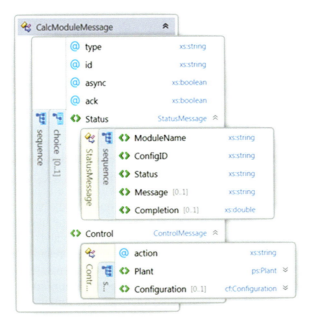

Figure 9.13. Structure of the Calculation module message

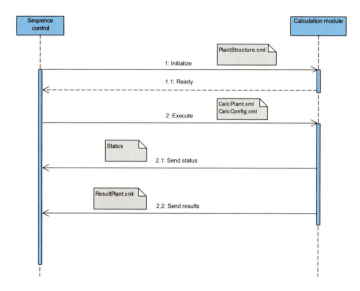

Figure 9.14. Sequence diagram of remote calculation protocol

Between sequence control and remote calculation modules a simple communication sequence has been defined, which is shown in Figure 9.14.

Initially, the sequence control component provides the plant structure model during invocation of the *Initialize-call*. This allows the calculation module to initialize its internal memory structures.

During execution of an evaluation cycle, the *Execute*-method of the calculation module is invoked passing a plant model containing all the data pertinent to the current evaluation run. During its computation, the calculation modules periodically call the Send *status* method to provide an update of its computation progress. After the computation is finished, the results are returned in the Send *results* method, again in the form of a plant model containing the computational results.

The remote invocations and replies are serialized as XML-formatted messages. Each message consists of either a plant model wrapped in a thin structure containing the invocation-related parameters (name of the method invoked, status flags) or a simple *status* message for the Send status method.

Within the EWave project, the remote module interface is used in particular by the developed optimization module, which runs on a Linux-based platform, while the sequence control module as part of the execution engine has been implemented using .NET Framework on Microsoft Windows. The distributed system approach has helped the project to separate development and deployment responsibilities as well as allow different runtime environments for separate components.

In the future, it is envisioned that decision support systems could be implemented as a highly-decoupled system of distributed microservices running in an industrial IoT cloud. The EWave system's remote calculation module approach provides a humble first step towards such a system.

9.3 The EWave sequence

Section 9.2 depicted all the main concepts of the EWave architecture, namely the data model, the general concept of the sequence of calculations and the calculation module interface. Thus the toolkit to build and execute the specific EWave sequence has been specified and shall be described in detail in this chapter.

9.3.1 Overview

The optimization of all control setpoints is the central piece of the whole EWave sequence. It computes results for the complete forecast horizon $\Delta T_{Horizon}$ in a resolution of small time steps Δ step. In EWave default values are chosen by $\Delta T_{Horizon} = 24$ hours to cover a whole day of operation and Δ step $= 15$ minutes to enable reasonable long time intervals between the switching of setpoints and small enough ones to enable efficient control. The optimization requires an initial state Z_{Init} of the whole water distribution system. This requires consistent values for all states, like flows, pressures, filling levels, setpoints, etc., which cannot be derived only from measurements. Thus, a simulation calculates that state by performing a dynamic simulation that starts at a past point in time t_{Past} that is dated back sufficiently and ends at a point in time t_{Init} at which the optimization starts. When defining the point in time t_{Init} the aspect of the duration ΔT_{Offset} of a whole EWave sequence and the resulting validity of its results has to be considered. To enable the simulation and especially the optimization algorithms to converge to reasonable results, the permitted ΔT_{Offset} of an EWave sequence has to be at least in the range of some minutes. Tests and results have shown that a ΔT_{Offset} of about 30 minutes is adequate to reach feasible results. That in turn implies that the point t_{Now} when an EWave sequence is started and the point the results are presented to the operator differ by ΔT_{Offset}, which leads to the necessary adjustment of $t_{Init} = t_{Now} + \Delta T_{Offset}$. To complete the main sequence of the EWave calculations an additional dynamic simulation on the interval $[t_{Init}; t_{Init} + \Delta T_{Horizon}]$ is performed taking the optimized setpoints into account. This simulation verifies the optimization results with a more detailed simulation model and yields system KPIs that are important for the operator. The single sequence of the main EWave modules and their placement in the time frame is shown in Figure 9.15.

9.3.2 Simulation state calculation

As stated, the available measurements are not sufficient enough to provide all required values for an initial state Z_{Init} from which the optimization or the simulation forecast can start their calculations. Nevertheless, the simulation takes into account as many measurements as possible to identify the system state at t_{Init} and performs an adaptive filling level tracing at least for the interval $[t_{Past}; t_{Now}]$ (see Chapter 4, Section 4.3.3 and Chapter 7, Section 7.4). Unfortunately, the most recent measurements have at best a timestamp close to t_{Now}. For simplicity, we assume to have measurements for t_{Now}. Thus, for the interval $[t_{Past}; t_{Now}]$, the measurements of the tank filling levels $[F_{Past}; F_{Now}]$, the control setpoints for pumps, valves and wells $[C_{Past}; C_{Now}]$, and

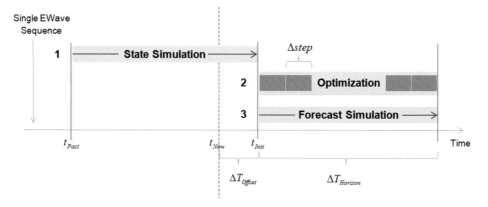

Figure 9.15. Single EWave sequence of main modules

the selected flows to derive the node demands $[D_{Past}; D_{Now}]$ are utilized. For the time interval $[t_{Now}; t_{Init}]$, the simulation has to perform a forecast simulation by using several assumptions:

- As ΔT_{Offset} is a small time interval of about 30 minutes, the demand (flow) values $]D_{Now}; D_{Init}]$ can be set constant to the value D_{Now}. A possible appliance of the demand forecast algorithm would not change the values significantly.
- As the simulation now performs a forecast, no adaption to filling levels is required. Thus measured filling levels do not have to be considered during that time interval.
- The control setpoints $]C_{Now}; C_{Init}]$ are extrapolated continuously from the last measured value C_{Now}. Their unknown course of action is predicted by taking real operation planning settings (see Section 9.3.6) and results from the last EWave optimization run into account as it is assumed that the operator controls the plant as EWave suggested to do.

Figure 9.16 shows the detailed data flow and time processing of the state estimation call to the simulation module.

The choice of the time-point t_{Past} depends on the overall EWave execution state as the dynamic simulation has to be granted a time interval $t_{Now} - t_{Past}$ that is long enough to adapt to the current measured filling levels. If the EWave system is booted up (cold start) there is no previously derived system state and t_{Past} has to be appointed some hours into the past. In contrast, for a running cyclic EWave system, each new state estimation simulation can already utilize states Z_i from the last cycle (warm start). Reasonable values for t_{Past} have been determined during EWave test runs and have been set empirically to $t_{Past} = t_{Now} - 12$ hours for the cold start and $t_{Past} = t_{Now} - 2$ hours for the warm start. Z_{Init} is the result of the state calculation and contains a consistent set of all flow and pressure values of all system components as well as the control setpoints of the pumps, valves and wells.

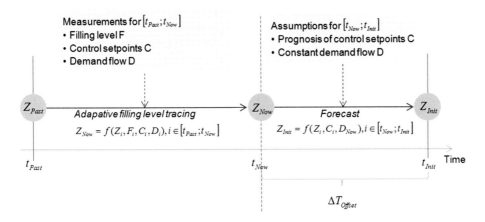

Figure 9.16. Detailed view of the state estimation simulation

9.3.3 Optimization

The optimization module calculates the optimized control setpoints O for the whole forecast horizon $\Delta T_{Horizon}$ in a resolution of Δ step. The optimization module has to know of all relevant system conditions that influence the result and thus requires different types of input:

- A consistent initial state Z_{Init} of the system at point in time t_{Init}, which is provided by the state simulation.
- Various types of boundary conditions, such as single maximum or minimum values for states and setpoints, more complex ones like water mixing ratios or even long-term rules for water volume production.
- Operational planning values of the operator and maintenance staff for $\Delta T_{Horizon}$ including e.g. maintenance service of a pump or the cleaning of a tank.
- Several prognoses for $\Delta T_{Horizon}$ including water demand D, self energy production P and electricity tariff T.
- A first initial state of all optimized setpoints O_{Init} for $\Delta T_{Horizon}$ to start the optimization from.

Specific boundary conditions merged with the operational planning values result in operational constraints for selected control setpoints C for the whole forecast horizon. A detailed insight in the preprocessing of the boundary conditions, the operational planning and the prognoses is outlined in Section 9.3.6.

Figure 9.17 sums up all the data flow and the time processing of the optimization module call.

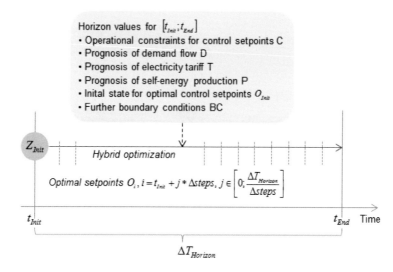

Figure 9.17. Detailed view of the optimization

9.3.4 Simulation forecast

The following forecast simulation (see Section 7.4) verifies the optimized setpoints with reference to feasibility, as its model is more detailed than the optimization model. Especially physical constraints are in focus, like tank levels – so tanks do not overflow or run dry – and pressure values – such that the pressure constraints at every system location are met. Additionally, the forecast can compute further key values the optimization did not consider in its objective function but the EWave users are still interested in. This includes further performance indicators of a water distribution system like outlined in Chapter 10, Section 10.1.

The simulation forecast applies the optimized control setpoints and considers the same prognoses values as the optimization (see Figure 9.18). It computes the course of all state variables over the whole time horizon $\Delta T_{Horizon}$ in a simulation specific output time step.

The results of the optimization and simulation module are data in quite raw format from the perspective of users like an operator or a manager. The post-processing of these results has to be done before displaying any of them. This is also outlined in Section 9.3.8.

The main EWave sequence and its calculation modules simulation and optimization have now been depicted. To guarantee that EWave is capable for deployment and execution in parallel to a real water distribution control, the necessary input and output data of the main calculation modules have to be provided continuously and robustly with consistent data. These additional modules with their underlying concepts and parts of their implementation are outlined in the following subchapters.

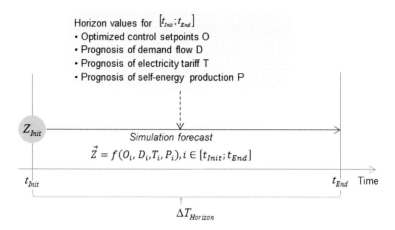

Figure 9.18. Detailed view of the forecast simulation

9.3.5 Measurement processing

Several calculations in EWave require the latest measured field data from the water distribution system. This includes e.g. filling levels of tanks for the state estimation, control setpoints for correct initial optimization values and flow values for the extrapolation of demands.

RWW stores its measured field data in data points in a database. EWave requires for every cyclic evaluation the newest extract of all specified data points in a processible format. For each data point the values of the last 24 hours (as this is required by the simulation state estimation) are saved in a csv file in a 1-minute resolution. As the EWave cycle is a multiple of 30 minutes, this export is executed every 5 minutes.

The necessary data points can be classified as follows:

- filling levels of all simulated tanks (except tank station Oberhausen-Tackenberg, as it is considered as demand node)
- current setpoints of all valves, whose opening degree is optimized
- flow measurements of all control valves, whose flow is optimized
- the measurements do not contain any setpoint values of pumps. Therefore, EWave imports the operating hour counter of each pump (regardless if a pump is fixed or speed controlled). The value in the RWW database hereby represents the percentage of an hour the pump was turned on in that resolution interval

$$\text{percentageOfHour} = \text{percentageOfOnInInterval} \times \frac{\Delta T_{ResolutionInterval}\,[s]}{3600\,[s]} \quad (9.1)$$

EWave Component	EWave Variable	EWave Unit	RWW Datapoint	RWW Unit	Conversion	EWave type
TankUeftermark	hAdapt	m	20001234	m³	/ area + z0	Equidistant TimeSeries
RWPump1HOL	onoff		30001234	h	h2onoff	TimeValuePair Profile
GalerieUEF	u		20004321	m³/h	/3600 / qNominalConst	Equidistant TimeSeries
Global	DemandHol	m³/s	20005678	m³/h	/3600 − 0.8 * DP20008765 / 3600	

Figure 9.19. RWW database mapping specification

For example if a pump is turned on 50% of the time of a given measurement resolution interval of 60 seconds, the value in the RWW database would be 0.0083.

- Flow values of the water sources (wells) as those are modeled only by their feed rate values.
- Flow at the outlet of the waterworks Dorsten-Holsterhausen as this value is used as baseline for the network demand prognosis and for the calculation of the current flow distribution on the demand nodes.
- Flow at the inlet of the Tackenberg tank unit, as this is used for correction of the network demand as well as the Tackenberg demand prognosis calculation.

The processing of those measurements contains the mapping of each data point to a variable in the EWave data model and necessary precalculations, like unit transformations or applying arithmetic operations including also other data model values or data points. The mapping functionality shall be designed to easily specify and extend the measurement relations. Thus the measurement import is specified in a processible table with the structure depicted in Figure 9.19 (including some example variables).

Figure 9.20 sums up the flow of the measurement processing from extracting data from the database up to inserting processed values into the data model. The EWave measurements processing module only provides values to the EWave model. It does not need any input data from the EWave model.

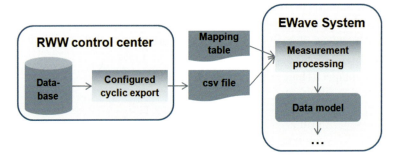

Figure 9.20. The measurement processing workflow

9.3.6 Processing of boundary conditions

When operating real water distribution systems, a lot of boundary conditions have to be considered by the operators when defining setpoints. There are different categories of such conditions.

- Operational constraints: These constraints are conditions that result directly in setting a fixed value or the limits of a control setpoint for a specific time interval. For example the flow operating range of a pump, direct presetting of control setpoints or the downtime of a pump.
- Tight boundary conditions: These conditions define required behavior of physical variables or setpoints within the forecast horizon $\Delta T_{Horizon}$ that is influenced by the current values of the control setpoints. Examples are minimal pressures, maximum flow volume, allowed or required combination of pumps and minimum up or down time of pumps.
- Long-term boundary conditions constraints: Water regulations often also include conditions that last longer than the forecast horizon of a current optimization run. Thus, a weighing of favorable and unfavorable compliance with these setpoints has to be made – leading up to fixed operational constraints when hitting the bounds of the conditions.

All of the above mentioned categories can be specified during the engineering of the water distribution network and set in the engineering plant model. But only a few of those conditions can be passed directly to the optimization as the evaluation of their values requires the consideration of the current and historic state of the system and additional operator input. Thus, other EWave modules take the engineered conditions into account and evaluate the necessary input values for the optimization.

Figure 9.21 gives an overview of the involved modules and the data flow for the condition management in EWave.

The EWave data model provides in general two possibilities to define boundary conditions. At first there are component-based variables. Boundary conditions that belong to single component instances can be defined directly in the component variables. The precondition is that they are part of the library specification. Those variables can be parameters, states or profiles and the optimization module has to transform these variables individually to the optimization model language. Examples for single parameters that are set in the engineering are the minimum down time of pumps or the maximum filling level of a tank. The minimum filling level of a tank already requires a more diverse handling. On the one hand there is a minimum level that has to be respected due to physical reasons or operational rules and thus can be set already in the engineering of the plant. On the other hand the optimization over such a long interval $\Delta T_{Horizon}$ would totally exploit it, if only such minimum filling levels would have to be considered. All tanks would be only filled at the end of $\Delta T_{Horizon}$ with their minimum required levels. Thus in EWave the required minimum filling levels for the time-point t_{End} at the end of the forecast horizon of each cyclic evaluation are set during runtime to be equal to the values at t_{Init}.

An even more complex example that is evaluated during runtime by the setpoint presetting module is the evaluation of a pump setpoint. Figure 9.22 shows two simple

172 T. Schenk et al.

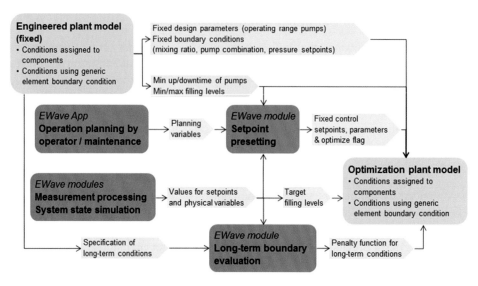

Figure 9.21. The processing of boundary conditions

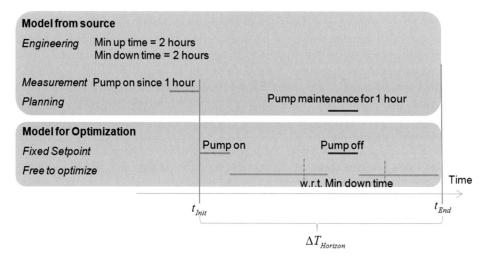

Figure 9.22. Evaluation of a pump setpoint input for the optimization

incidents that have to be taken into account. A pump already running at the beginning of an optimization run has to be running till at least reaching its minimum up time interval. A pump scheduled for maintenance cannot be used for optimization in that time interval and additionally the minimum down time interval has to be considered by the optimizer as well. Thus a separate pump variable indicates that the pump is not free to be optimized in that time intervals.

The second possibility to specify boundary conditions is to use a generic boundary condition element. This element is used to specify most of the tight boundary

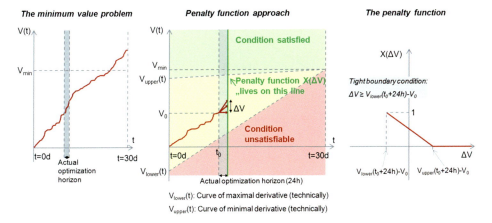

Figure 9.23. The penalty function for long-term boundary conditions

conditions and long-term boundary conditions. Its structure provides the ability to specify for arbitrary variables of multiple instances of components simple arithmetic equations and inequations, like weighed linear combinations or products including variable offsets. This type of boundary condition also can contain all specifications that are necessary to define long-term boundary conditions. Goal of the module for long-term boundary evaluation is to provide the optimization with a penalty function, whose output has been standardized to the interval $[0, 1]$. The optimization then can integrate a modified penalty function into its cost function as it suits the optimization algorithms, e.g. requirements may be smoothness in general and a rising, high gradient in regions where the boundary condition is likely to fail. Figure 9.23 shows an example of a long-term boundary condition for a minimum water volume that has to be pumped from a well in a 30 days period. The penalty function is built for each optimization run based on the value V_0 of the target variable (here volume) and the curves (here fixed lines) of the minimum V_{Upper} and maximum V_{Lower} gradient of the target variable (these can be mostly derived technically).

The penalty function $X(\Delta V)$ can then be computed as

$$X(\Delta V) = \begin{cases} \frac{V_{Upper}(t_0+24h)-V_0-\Delta V}{V_{Upper}(t_0+24h)-V_{Lower}(t_0+24h)}, & \text{(if } \Delta V < V_{Upper}(t_0 + 24h) - V_0) \\ 0, & \text{(else)} \end{cases} \quad (9.2)$$

For maximum problems, the penalty function is set up likewise just inverted.

This approach has been applied for patent under *2015E18143 DE: Efficient optimal short-term control of water supply networks comprising long-term regulatory and economic targets.*

Recalling Figure 9.21, the modules for measurement processing and state calculation as well as the long-term boundary condition and at least parts of the setpoint presetting have been described in detail. It remains to explain how EWave specifies and receives the operational planning variables.

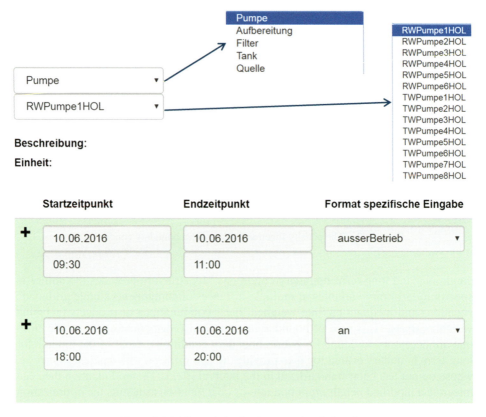

Figure 9.24. The website for the operational planning

The operation planning module allows EWave to consider the operational decisions made by the staff, e.g. cleaning a tank, maintaining a pump, or presetting the flow from a well to a fixed value for a specific time interval. EWave has to get notified of such events, otherwise those variables would be free to be modified by the optimization algorithm and its resulting setpoints are not applicable to the real system. The operation planning application has two aspects, the user interface and the transformation to the EWave data model.

The entry to such operational planning is possible for the operators at any time. So the application has been realized as a separate program running as a client-server application. This way, it would be even possible to use mobile devices in the intranet of the water supply company for the specification of the planning intervals. The website (Figure 9.24) offers the possibility to enter for a specific element a start time and an end time for a specific operational planning action.

The possible actions that can be specified on the user interface have been derived from discussion with the water supply company RWW and cover the most relevant actions the operator would take. Thus, these actions are specified from an operator point of view. There are the five clusters pumps, tanks, water sources, filters and

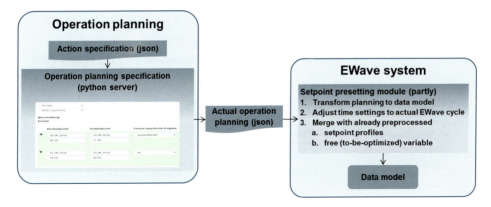

Figure 9.25. Process of the operational planning

treatment. Pumps can be switched on or off. Tanks can be set out of operation in the three different ways, namely to be empty, to hold a specific filling level and without any further specification. The treatment elements (single UV stations, whole UV units or degasifiers) can be set out of operation. For the filters, a whole filter line can be set out of operation or each single filter individually. Additionally each single filter can be set to be flushed. The infeed Üfter Mark can also be set to be out of operation and each of the wells can be set to fixed flow rates.

The second aspect of the operation planning is the transformation of those above named operator-oriented actions to the EWave library and data model. Many of the transformations are 1:1 mappings like the on/off switching of a pump. But there are also some complex transformations that have to be managed. One example is the unavailability of a filter, since this changes the synchronization control of the flow in both filter lines and thus the boundary condition. Another one is the cleaning of a tank, which corresponds to the UI action of an unavailable tank with zero filling level. This implies a chain of valve actions: firstly a parallel closing of the valve in front of the tank that should be emptied and of the valve behind any parallel tank (as otherwise there would be a flow back to the tank that should be emptied), secondly – when the tank is empty – closing the valve behind the empty tank and opening the valves behind the other parallel tanks. These proprietary control chains of some of the actions have made it quite hard to specify the transformations in a generic format and process them automatically. Thus the transformations have been implemented in separate functions of the setpoint presetting module (Figure 9.25).

9.3.7 Demand prognosis

To complete the data preprocessing setup for the optimization, the prognosis data needs to be provided. For the water demand prognosis, a separate EWave demand prognosis module calculates for each demand node of the water distribution system its demand flow for the whole forecast horizon.

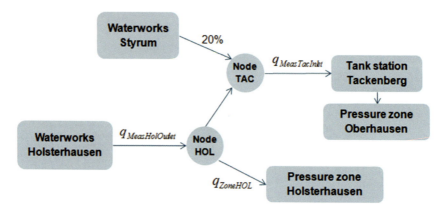

Figure 9.26. Flow distribution in the water distribution network

The water distribution network in EWave consists of more than 50 demand nodes realized as virtual tanks, but demand prognosis evaluation is not done separately for all these nodes. The concept in EWave is to provide a demand prognosis for only a small number of nodes, each node representing a pressure zone. The water demand is then distributed by a simple percentage distribution to all the other related nodes of that pressure zone (see Chapter 6, Section 6.2.7). For the Holsterhausen network there are only two flows for which a demand prognosis evaluation is provided:

- The percentage of outlet flow of the waterworks Dorsten-Holsterhausen that serves the Holsterhausen pressure zone
- The inlet flow of the Oberhausen-Tackenberg tank station

Chapter 3, Section 3.2.5 already described a demand prognosis algorithm, which is implemented in a separate Matlab software program. This program calculates the demand at the Holsterhausen outlet node for the Holsterhausen pressure zone, which contains all demand nodes except the Tackenberg node. The program requires the measured value of the demand flow that has to be prognosed. The only available measurements are the complete outlet flow of the waterworks Dorsten-Holsterhausen and the inlet flow to the tank station Oberhausen-Tackenberg. Tackenberg receives water from Holsterhausen and additionally from the waterworks Mühlheim-Styrum Ost, whose portion can be assumed to be a fixed value of 20% of the whole inlet flow (see Figure 9.26). Thus the partial outlet flow for the demand prognosis of the Holsterhausen pressure zone calculates as follows:

$$q_{HOL} = q_{MeasHolOutlet} - 0.8 * q_{MeasTacInlet} \tag{9.3}$$

To be robust to measurement and abrupt demand fluctuation, q_{HOL} is averaged over the last hour.

After calculation of the demand prognosis profile, the percentage distribution table is applied and the demand prognosis for all virtual nodes of the Holsterhausen pressure zone is set accordingly (see Figure 9.27).

Component	Variable	Unit	ReferenceDemand	Unit	Distribution
VirtualTank148	qExt	m³/s	DemandHOL	m³/s	-0,061292841
VirtualTank136	qExt	m³/s	DemandHOL	m³/s	-0,000634906
VirtualTank109	qExt	m³/s	DemandHOL	m³/s	-0,061292841
VirtualTank135	qExt	m³/s	DemandHOL	m³/s	-0,047386682

Figure 9.27. Excerpt from the Holsterhausen demand distribution

The Tackenberg tank station is filled manually every night in 4–6 hours to serve the demand of the pressure zone Oberhausen the whole following day. As the volume sums up to only some percent of the whole waterworks Dorsten-Holsterhausen outlet, EWave implements here a trivial approach to provide a demand prognosis for the inlet flow of the Tackenberg tank station: The measured inlet flow of the last 24 hours is just repeated for the prognosis horizon.

9.3.8 Cyclic evaluation

So far, the different modules of a single EWave sequence were explained in detail. Before describing the last module of the sequence, the evaluation of switch message, the aspect of cyclic evaluations should be understood.

The complete single EWave sequence is automatically repeated at equal time intervals. The time horizons of the different modules of the sequence are shifted simultaneously. A cyclic evaluation with receding horizon is the consequence.

The time interval for the automatic re-calculation ΔT_{Cycle} is chosen small compared to the time horizon of the optimization in order to allow for a quick response to unforeseen dynamics in the loads and boundary conditions. But since loads and boundary conditions usually change gradually, the results from the preceding evaluation cycle are still valid to some reasonable extent and should be re-used in the current evaluation as will be explained below. For an easy integration of the preceding results, ΔT_{Cycle} is required to be an integer multiple of the time step Δ used by the simulation and optimization modules but they do not need to be equal. Obviously, one should further require $\Delta T_{Cycle} \geq \Delta T_{Offset}$, otherwise the evaluation starts before the preceding results even get used. In the pilot implementation, $\Delta T_{Cycle} = \Delta T_{Offset}$ is used. For simplicity, we limit ourselves to this case in the following. The cyclic evaluation and its time aspects are sketched in Figure 9.28.

The optimization results from the preceding evaluation cycle are re-used in two ways

- In the simulation-based state estimation, there is a gap between the point of time t_{Meas} of the latest measurements available for evaluation and the endpoint for simulation t_{Init}. For this period of time, operator actions as suggested by the preceding optimization(s) are assumed.
- For the optimization, the operation schedules from the preceding evaluation extrapolated to the new end of optimization horizon are used as initial guess for the new schedules. Since they were optimal in the latest time horizon, they

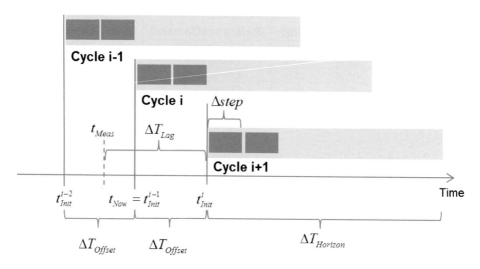

Figure 9.28. Cyclic evaluation with receding horizon

will still be not too far away from optimality in the current horizon. By doing this, computational time is reduced significantly.

The way how optimized schedules are translated into precise operator actions, how these instructions are updated due to the cyclic evaluations and aligned with latest measurements is a complex topic in itself and will be discussed in the next paragraph.

9.3.9 Evaluation of switch messages

The EWave UI shall enable primarily the operator to know when to set which setpoint value. But results from the optimization and simulation cover vast calculation results in different formats and types. To manage this complexity, EWave introduces a message board (described in detail in Chapter 10, Section 10.2.3) that enables the operator to easily recognize which actions to take at present and in the very near future. Thus, a further EWave module transforms measurements and optimization results to a format usable by the message board.

At first, the setpoints are divided into two groups:
- continuous setpoints like pressure, flow or speed/opening setpoints for pumps, valves and wells and
- discrete setpoints like pump or well on/off switches.

Secondly, the display of the optimized setpoints has to consider that operators cannot immediately change all setpoints to the proposed values at once at the recommended point in time. Operators can be absent for some times or have other duties, setting values in the control center also takes some time and the sheer number of values to change can lead to delays as well.

For **continous setpoints**, the optimization probably produces different values for each future time interval Δstep over the whole time horizon $\Delta T_{Horizon}$ (see Chapter 9, Section 9.3.3), which would result in a confusing amount of actions for the operator at each time interval. Additionally, the relative difference between the actual measured setpoints and those optimized values – as they are continuous numbers evaluated by a numerical optimization – might be considered as too small by an experienced operator to have enough considerable effect. Thus, EWave displays for continuous setpoints only the **actual** measurement value and the **actual** optimized value at the present point in time. The operator can then decide if it is worth to undergo the process of changing the setpoint in the control center. A future implementation could evaluate this automatically taking type of setpoint and affected system component as well as absolute and relative difference to the actual measurement and further aspects into account.

Discrete setpoints have to be treated differently as they have mostly a huge impact on the physical system and require disruptive and mostly promptly action. In EWave this led to a new type of time profile holding punctual events, namely the switching times of the pumps with an integer value of $[-1; 1]$. A value of -1 indicates switching off and a value 1 switching on. As a consequence also an empty profile of that type holds information as it states that no switching is necessary. This type of profile is processible by the message board as each event is directly displayed there.

The EWave evaluation of the switching points considers that there is a significant time lag ΔT_{Lag} between the most actual measurement of setpoint values at t_{Meas} and the point in time t_{Init} when the optimized setpoint values of the actual calculation is becoming valid (see Figure 9.15), with $\Delta T_{Lag} > \Delta T_{Offset}$. Let us consider the actual evaluation cycle (consisting of a whole EWave sequence) starting at t_{Now} to be cycle i, then, there have been previous EWave cycles $i - 1$ and $i - 2$. The course of action in time is depicted in Figure 9.29.

The requirements for the switching instructions are:

- Taking the time lag ΔT_{Lag} into account the switching instructions should be as actual and as correct as possible.

- One and the same switching instruction should not be displayed on the message board for different points in time (e.g. for t_{Init}^{i-2} and t_{Init}^{i-1} but only once for the earliest displayed occurrence (in the given example for t_{Init}^{i-2} then).

- Past switching instructions should be still displayed but only date back one cycle interval ($t_{Now} - \Delta T_{Offset}$). Thus the message board enables a consistent and transparent display for the operator.

- Switching instructions that date back more than two cycles (and would drop therefore from the message board) and have not been followed yet and are still proposed by the optimization should show up again. This way no instruction is lost due to some possible absence or other duty of the operator.

To fulfill these requirements, two evaluations of switching instructions are inserted into the EWave sequence (s. Figure 9.29): One close to the start of a new evaluation

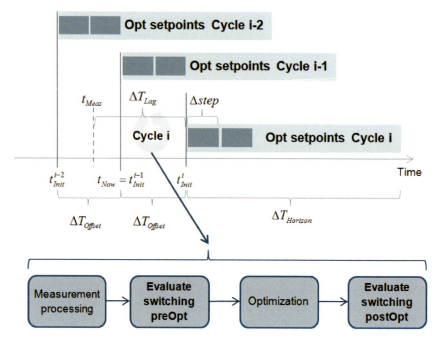

Figure 9.29. Relevant time considerations for the switching evaluation

cycle directly after the processing of the measurements of point in time t_{Meas} and another directly after the optimization run.

The switching evaluation **before** the optimization leaves all switching instructions from the last cycle $i - 1$ as they are, except that it will insert a new switching setpoint for t_{Now}, if the latest measured setpoint value from t_{Meas} differs from the optimized setpoint value from t_{Init}^{i-1} (of cycle $i - 1$) **AND** if there is no switching instructions present at t_{Init}^{i-1} or even at t_{Init}^{i-2} (to avoid multiple display of the same switching).

The switching evaluation **after** the optimization

- leaves instructions for t_{Init}^{i-2} and $t_{Init}^{i-1} = t_{Now}$ as they are,
- deletes all future switching instructions (from t_{Init}^{i} on),
- sets a switching instruction for t_{Init}^{i}, if the optimized setpoint of t_{Init}^{i} differs from the expected setpoint value at t_{Init}^{i} (the expected setpoint value assumes that the operator followed the switching instructions from the last cycle $i - 1$),
- and sets future switching instructions (from t_{Init}^{i} on) for a point in time $t_{Init}^{i+j} * \Delta step$, $j \in [1; \frac{\Delta T_{Horizon}}{\Delta step} - 1]$, if the two neighboring optimized setpoints of the time point differ.

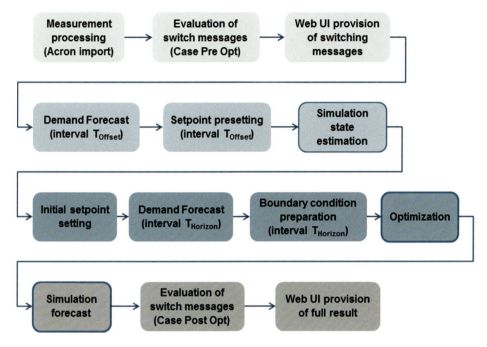

Figure 9.30. Summary of single EWave cycle

9.3.10 Summary

All evaluations and modules described in this chapter are called in one single EWave evaluation (or so called EWave cycle). The sequential course of action consists of the four main parts measurement evaluation, simulation state estimation, optimization and presentation with all the necessary pre- and post-calculations (see Figure 9.30).

Bibliography

[1] Amesim, https://www.plm.automation.siemens.com/de/products/lms/imagine-lab/amesim/index.shtml.

[2] M. Andersson, *Object-Oriented Modeling and Simulation of Hybrid Systems*, Department of Automatic Control, Lund Institute of Technology (LTH), 1994.

[3] J. Fischer, B. Obst, and B. Lee, *Integrating material flow simulation tools in a service-oriented industrial context*, 2017 IEEE 15th International Conference on Industrial Informatics (INDIN), IEEE, 2017, 1135–1140.

[4] HEEDS, https://www.redcedartech.com/index.php/solutions/heeds-software.

[5] P. Hehenberger and D. Bradley, Digital Twin – The Simulation Aspect. In *Mechatronic Futures Challenges and Solutions for Mechatronic Systems and their Designers* (S. Boschert and R. Rosen, eds.), Springer, 2016, 59–74.

[6] Matlab. https://de.mathworks.com/.

[7] S. Mattsson and H. Elmqvist, *Modelica – an international effort to design the next generation modeling language*, 7th IFAC Symp. on Computer Aided Control Systems Design, CACSD'97, 1997.

[8] Microservices, http://microservices.io/patterns/microservices.html.

[9] Model Center, https://www.phoenix-int.com/.

[10] Modelica Association, https://www.modelica.org/

[11] T. Schenk, A. Gilg, M. Mühlbauer, R. Rosen, and J. C. Wehrstedt, Architecture for modeling and simulation of technical systems along their lifecycle. *Computing and Visualization in Science* 17(4) (2016), 167–183.

[12] *Single source of truth*, https://en.wikipedia.org/wiki/Single_source_of_truth.

[13] ZeroMQ, http://zeromq.org/.

Chapter 10

Water cockpit: dashboards for decision support systems

Michael Plath, Constantin Blanck, Stefan Fischer, Moritz Allmaras, Andreas Pirsing, Tim Schenk, and Annelie Sohr

Abstract. In this chapter, it is discussed which indicators and views of water supply systems are needed to address all relevant user roles targeted by a decision support system. The first step is an overview of the indicators that are used to estimate and compare the efficiency of water supply systems. The specific energy consumption is only of limited usefulness in this context, hence hydraulic and electric efficiency indicators are defined that allow the formulation of an overall efficiency of a plant. Then, the relevant user roles and related quality attributes for a decision support system are investigated. The chapter concludes with a detailed discussion on how the user interface for the EWave system has been designed and implemented in order to meet these system qualities.

10.1 Energy and process

10.1.1 Industry-specific indicators, especially efficiencies

For the evaluation and comparison of energy efficiency and ecological sustainability of various plants and waterworks, key performance indicators (KPI) were defined in the EWave project.

When assessing the energy efficiency of water supply systems, the specific energy requirement in kWh/m^3 is usually used. For different extraction, processing and distribution situations, this value differs depending on the corresponding framework and boundary conditions. In Germany, the specific energy requirement ranges from $0.18\,kWh/m^3$ to $0.96\,kWh/m^3$ for the entire process from water production to distribution [6].

A list of the currently used indicators, mainly the specific energy requirements for the different levels of water supply, was compiled on the basis of relevant literature [11]. Starting with a higher-level corporate KPI up to a plant-specific KPI. From the application of the KPI it is known, however, that the boundary conditions have a strong influence, so that an evaluation on the basis of these KPI is difficult. The specific energy requirement is not sufficient for an evaluation and comparison of different plants and plant components. One example is the specific energy requirements of pumps. Depending on the operating point and thus depending on the flow rate, this can vary greatly. This is why the efficiency of pumps is usually given as kWh per m^3 and per m discharge head. This is possible for pumps, but not for all other system components. The approach was now to create the possibility to compare plants and plant components with each other by means of the industry-specific KPI. It was also important for RWW that the KPI could be used in the energy management system and thus help the plant operator directly.

To be able to compare plants and plant components with each other, unitless efficiencies, which can have values between 0 and 1, must be defined. As a result of the development, it should be possible to consider electrical and hydraulic efficiencies together. The subsequent measures to improve the respective efficiency and energy savings must then be analyzed in detail. In the following, the calculation rules for the different efficiencies are presented.

Electrical efficiency. Electrical efficiency can be used when electrical energy is converted into another form of energy, e.g., electricity into hydraulic energy in pumps. Electrical efficiency is defined as the ratio of mechanical power output to electrical power input. In literature, the calculation of the electrical efficiency is often described with different symbols. The most common abbreviations are shown in Equation (10.1).

$$\eta = \frac{W_{out}}{W_{in}} = \frac{\text{output work}}{\text{input work}} = \frac{P_{out}}{P_{in}} = \frac{\text{output power}}{\text{input power}} \quad (10.1)$$

A classic example is the calculation of an electric motor. A mechanical nominal power is output at the motor shaft at a defined electrical work input. The difference between the input and output power is defined as loss. The ratio of work input to work output results in the specific motor efficiency. For a downstream gearbox, which also has a specific efficiency, the overall efficiency is calculated by multiplying both individual efficiencies.

The efficiency of a pump unit consists of the electrical efficiency of the motor and the hydraulic efficiency of the pump. The definition of the efficiency of a pump unit is shown in Equation (10.2).

$$\eta_{el.\ pump} = \frac{W_Q}{W_{el}} = \frac{\text{conveying or hydraulic work}}{\text{electric work}} \quad (10.2)$$

The electrical work of the pump unit is measured, the effective one must be calculated on the basis of the available measurement data. In addition to the flow rate Q, the discharge head H and the geodetic heights are required for the calculation of the effective work.

The Equation (10.2) refers only to the calculation of the electrical efficiency of pumps. Equation (10.3) shows general formula valid for all electrical systems, e.g., also for UV-reactors.

$$\eta = \frac{W_{eff}}{W_{el}} = \frac{\text{effective work}}{\text{electric work}} \quad (10.3)$$

The Equation (10.4) shows the calculation of the overall electrical efficiency.

$$eta = \frac{\sum W_{eff}}{\sum W_{el}} \qquad (10.4)$$

Hydraulic efficiency. The hydraulic efficiency is to be used to describe the efficiency of all parts of the plant through which water flows. This allows the pressure losses, which are among other things dependent on the volume flow, to be evaluated. The pressure energy present in the system or in the pipeline is required to calculate the hydraulic efficiency. The pressure energy consists of the data of the volume flow, the pressure and the geodetic heights, analogous to the effective power, and is shown in Equation (10.5).

$$E_{p,n} = (Qt)(p_n h_n - h_{ref}\rho g) \qquad (10.5)$$

The Equation (10.6) shows the calculation of the hydraulic efficiency.

$$\eta_{hyd} = \frac{E_{p,end} - E_{p,ref}}{E_{p,start} - E_{p,ref}} \qquad (10.6)$$

The denominator in Equation (10.6) with the calculated pressure energy at the beginning of the considered section influences the result significantly, this means that the operating pressure has a decisive influence here. A pressure loss of 0.5 bar, for example, is rated more strongly in hydraulic efficiency at an operating pressure of 1.0 bar and less at an operating pressure of 5.0 bar. This must be taken into account when assessing the efficiency with the aid of hydraulic efficiency. This is why the pressure loss must also be taken into account for the evaluation.

The overall hydraulic efficiency is calculated from the pressure energy at the end of the observed section in relation to the sum of the pumping work supplied by the pumps. The calculation is shown in Equation (10.7). The pressure energy ($E_{p,reference}$) is equal to the pressure energy at the beginning of the system observed.

$$\eta_{hyd,tot} = \frac{E_{p,end} - E_{p,ref}}{\sum W_Q} \qquad (10.7)$$

Due to the hydraulic efficiency, it is possible to compare the pressure loss, caused by a change in the volume flow or other mechanical flow conditions. The calculation of the hydraulic efficiency is always applicable. Even with rising and then falling pipes with a pressure gain, the calculation delivers meaningful and correct efficiencies that only evaluate the pressure loss.

Plant efficiency. To calculate the plant efficiency, the previously defined efficiencies (electrical and hydraulic) must be combined. In this way, the plant efficiency evaluates system through which water flows, i.e., which have a hydraulic efficiency, but also consume electricity.

$$\eta = \frac{W_{eff} + E_{p,end} - E_{p,ref}}{W_{el} + E_{p,start} - E_{p,ref}} \qquad (10.8)$$

Overall efficiency. The overall efficiency of the waterworks can then be calculated on the basis of the overall electrical efficiency and the overall hydraulic efficiency. The sum of the electrical energy used in the plants represents the effort and thus the denominator of Equation (10.9). In addition to the effective work, the hydraulic work must also be taken into account in the meter. In order to include the hydraulics correctly, the pressure energy present at the end must be taken as a benefit and not the pumping work of the pumps (W_Q). Thus, the conveying work must be deducted from the sum of the effective work of the plant (W_{eff}).

$$\eta_{tot} = \frac{\sum W_{eff} - \sum W_Q + E_{p,end} - E_{p,ref}}{\sum W_{el}} \qquad (10.9)$$

10.1.2 Application test of efficiencies

For the Dorsten-Holsterhausen waterworks, not all data required for the efficiency calculations were yet available. Particularly the pressures required for the calculations were not yet available, but the measurements have since been added.

In order to check the calculations, detailed data from a measurement campaign at the Essen-Kettwig waterworks were used. The volume flow was gradually increased over several days and the pressure was additionally recorded at several points with mobile pressure loggers.

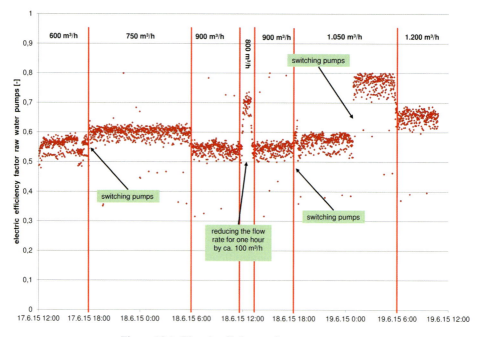

Figure 10.1. Electric efficiency of raw water pumps

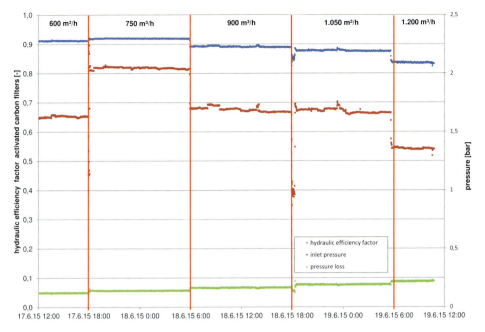

Figure 10.2. Hydraulic efficiency of the activated carbon filters incl. piping

Figure 10.1 shows the electrical efficiency of the raw water pumps. Jumps in efficiency can be seen in the volume flow adjustments and when pumps have been switched on.

Figure 10.2 shows the hydraulic efficiency of the activated carbon filters including a section of the pipeline. Changes can be seen here in the volume flow changes, the pressure loss increases with increasing volume flow. As the inlet pressure changes, this also affects the efficiency, as has already been described.

The efficiency of the ozone system is shown in Figure 10.3. This is mainly influenced by the electrical efficiency. The hydraulic efficiency component has only a small influence. Changes in efficiency can be seen in the volume flow increases and in ozone switches. In principle, it should be noted that the efficiency of the plants continues to increase with increasing treatment volume flow.

Finally, Figure 10.4 shows the overall efficiency of the waterworks. The efficiency is highest at the highest treatment volume flow set.

In order to calculate these efficiencies, additional, temporary measurements were necessary in some cases. In practice, these measurements can also be permanently installed and the data or the calculated efficiencies can be transferred to the control system. In this way, the control center staff could be provided with a benchmark that would make the efficiency-related effects of their switching actions visible. However, this implementation makes more sense for individual plant components, since a real-time view of the overall efficiency would require that no further switches are made during the reaction time of the entire system.

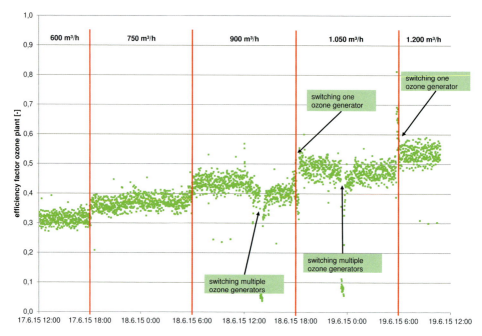

Figure 10.3. Efficiency of the ozone plant

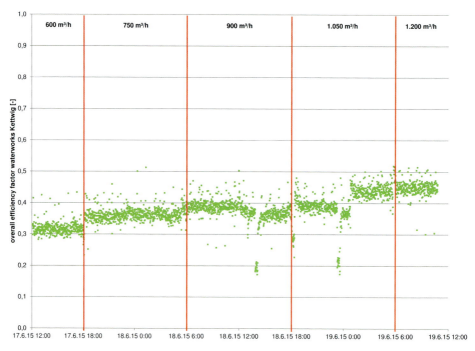

Figure 10.4. Overall efficiency factor of the waterworks

10.2 User interface technologies

10.2.1 User roles

Decision support systems (DSS) for water network operation address a range of different user roles in the organization of a water utilities company:

- Operators are responsible for the safe and reliable plant operation following the operational policies set by the company's engineering department.
- Engineers have the task to define, analyze and improve operational policies as well as plan and supervise installation and maintenance tasks.
- Managers oversee the business aspects of plant operation. They also have the overall economic efficiency of the organization under their supervision.

Operators. First and foremost, decision support systems need to address the needs of operators and support their day-to-day decisions on how to actuate the controllable assets. Operators often do not have a deep theoretical understanding of the inner workings of the water supply system but have extensive experience on what the short-term implications of certain control actions are. Hence, it is crucial that the DSS displays control suggestions in a way that is as close as possible to the representation of the plant that operators are used to.

Which assets can be controlled and what aspects can be addressed by inputting setpoints is varying widely between different utilities companies. Some have mostly automated the operation of the entire supply system, for others, every single pump needs to be operated manually by shift operation teams. Even for controllable assets, there exist many different variants on what aspects can be controlled: Pumps can be operated by an on/off switch or by providing a setpoint for flow or pressure.

The user interfaces to Decision Support Systems need to be easily customizable to address the individual control inputs that operators of a specific utility company are presented within their automation system.

Engineers. Engineers use decision support systems to plan for future operation of the water network. They ensure that the model used by the DSS during online operation is properly calibrated to match the behavior of the real system as good as possible. In addition, engineers are interested in analyzing specific scenarios, such as tank maintenance, firefighting or other emergency cases.

Managers. Managers are interested in the values of certain key performance indicators that quantify overall efficiency and other global economic properties of the plant operation. These values need to be presented in ways that do not require knowledge of the actual plant structure or the hydraulic functioning of the plant. Managers also like to have an indication of the cost saving due to the operation of the DSS itself.

10.2.2 System qualities & decisions

System qualities. The primary role that the EWave DSS addresses is the operator role. Hence, the nonfunctional requirements for development of the EWave system are related to issues relevant to the operators of water networks:

- Accessiblity
  The DSS system needs to be easily accessible to operators and may not distract them from their main responsibilities. The amount of training and specialized technical expertise required to operate the system needs to be kept to a minimum. The system should operate as autonomously as possible and present results in a way that is comprehensible to water network operators.

- Usability
  The user interface of the EWave system needs to be easily understood by water network operators and should be intuitive to use. It should include only the part out of the system that is relevant to the targeted user role.

- Verifyability
  Results calculated by the DSS should be easy to verify for plausibility. Hence, the system should not only output optimal operational schedules but also indicate the predicted system behavior that it assumed would enfold when these schedules are put into operation. This way, it can easily be tracked what decisions the systems suggests are based on which assumptions.

Decisions. Based on the identified system qualities, it was decided to implement a web-based user interface for the EWave system that can be run in any standard web browser. It is assumed that the standard operational environment for operators would mainly consist of workstation computers with desktop monitors, so the user interface does not primarily focus on supporting portable web-enabled devices such as smartphones or tablet computers. A minimum screen resolution of 1024×768 pixels is assumed.

The user interface should display a schematic view of the plant/network that is addressed by the DSS so that any inputs and results can be easily attributed to assets that the water network operator is familiar with.

10.2.3 User interface

The user interface (UI) of the EWave system operates in two different modes:

- In **operator mode**, the switch actions calculated by the EWave system are presented in a way that directly relates to the tasks of the plant operator. A fixed subset of the calculated simulation results that may be of interest to operators is available for visualization in charts.

- In **expert mode**, all calculation results are accessible, and results from past cycles can be retrieved and visualized as well. This mode is particularly suitable to engineers for understanding and validating the results generated by the EWave system.

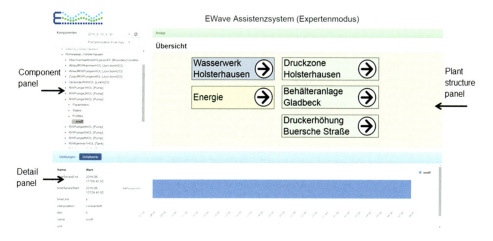

Figure 10.5. EWave user interface (expert mode)

While the operator mode targets operators as users of the EWave system, the expert mode addresses engineers with advanced knowledge of the plant hydraulics and the functioning of the EWave system. Conceptually, expert mode is used during setup and piloting of the EWave system to verify the calculation results and to ensure that the system operates reliably.

Figure 10.5 shows a screenshot of the EWave user interface in expert mode. The user interface consists of three main panels:

- The **plant structure panel** contains a visual representation of a part of the plant or water network.
- The **component panel** contains an expandable tree of the assets that are part of the plant.
- The **details panel** has two different tabs:
  - In the message tab, it shows a table of control actions that the EWave system suggests the operator to implement.
  - In the property tab, it displays properties of the component that is selected in the component panel.

In expert mode, the component panel contains two additional dropdown items for browsing results from historic calculations. Also, in the component panel, all variables are available for each component, while in operator mode only certain predefined variables can be selected.

Plant structure panel. As the water production and distribution grid covered by the DSS can become quite complex, the entire system is not displayed on a single page but distributed to multiple pages. In order to make navigation between these pages as simple as possible, there are specific navigation sheets that do not contain

192   M. Plath et al.

Figure 10.6. Overview navigation sheet (screenshot)

any components but only allow navigation between the different sheets of the plant structure view (see Figure 10.6).

Navigation sheets contain link icons that can be clicked to jump to the respective sheet inside the plant structure panel.

For pages that show an actual part of the plant, two different views have been developed:

- A process view that displays a symbolic representation of the plant components and their topological connections in a style similar to a *Process and Instrumentation Diagram* (P&ID, see Figure 10.7). This view is particularly suitable for process parts (e.g. water treatment facility) of the water plant.

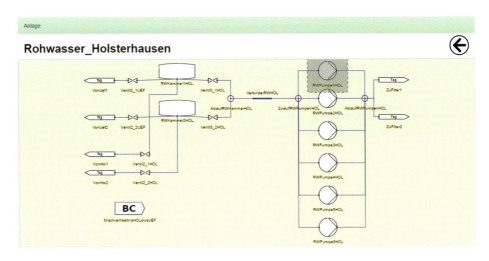

Figure 10.7. Process view (screenshot)

10 Water cockpit: dashboards for decision support systems   193

Figure 10.8. Geographic view with level of detail display

- A geographic view showing geographic location of plant components on a street map comparable to a *Geographic Information System* (GIS, see Figure 10.8). This view is particularly suitable for visualizing parts of the water system that cover extended geographic areas (e.g. distribution zones). The geographic view features a *level of detail* (LOD) approach that shows individual components only if their icons are spaced sufficiently far apart from each other according to the current zoom level. Icons that overlap are collapsed into a box that indicates only the count of hidden icons. This ensures that the geographic view remains clutter-free even in areas where many components are located geographically close to each other, but still providing a visual indication of where components of the plant are located even at overview zoom levels.

Both types of views allow for interactive drag and zoom behavior and as well as a selection of individual components by a single click.

In the top right corner of each sheet shown in the plant structure panel (except the initial overview sheet), a link icon is displayed that on click navigates to the navigation sheet that contains the link to the current sheets.

In addition, plant structure sheets of P&ID or GIS type support interaction with so-called tag components. Such tags symbolize hydraulic connections between components on different sheets and always occur in pairs of corresponding tags (see Figure 10.9). A single click on a tag component will navigate to the component (and corresponding sheet) that is connected to the counterpart of the tag.

194   M. Plath et al.

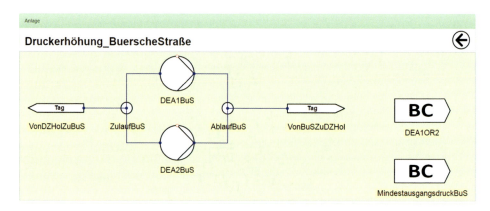

Figure 10.9. PID sheet with tag components (screenshot)

Component panel.   The component panel shows a collapsible tree of components of the water plant, structured by the pages that are shown in the plant structure view. On unfolding of the component items, details about their attached model variables (parameters, states and profile) are shown and can be selected (see Figure 10.10).

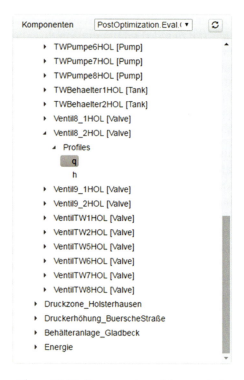

Figure 10.10. Component panel (screenshot)

10 Water cockpit: dashboards for decision support systems

	Zeitpunkt	Schaltempfehlung	Komponente	Ist-Wert	Soll-Wert
Zuvor	15.01.2017, 09:45	Pumpe einschalten	TWPumpe1HOL		
	15.01.2017, 09:45	Pumpe ausschalten	RWPumpe3HOL		
	15.01.2017, 09:45	Pumpe einschalten	RWPumpe2HOL		
Aktuell	15.01.2017, 10:00	Pumpe einschalten	TWPumpe2HOL		
	15.01.2017, 10:00	Mengenziel Aufbereitung Holsterhausen setzen	Ventil5_1HOL	0.40354	0.4486
	15.01.2017, 10:00	Mengenziel Aufbereitung Holsterhausen setzen	Ventil5_2HOL	0.40365	0.4486
Demnächst	15.01.2017, 10:00	Pumpe einschalten	TWPumpe1HOL		
	15.01.2017, 10:15	Pumpe ausschalten	RWPumpe3HOL		
	15.01.2017, 10:15	Mengenziel Aufbereitung Holsterhausen setzen	Ventil5_1HOL	0.43983	0.4346
	15.01.2017, 10:15	Mengenziel Aufbereitung Holsterhausen setzen	Ventil5_2HOL	0.44034	0.4346

Figure 10.11. Message tab (screenshot)

Details panel. The details panel contains a tabular view with two tabs for messages and values. The **message tab** shows a table of switch actions that are recommended by the EWave DSS (see Figure 10.11). All entries in the table are derived from the last cyclic run of the EWave system. Switch actions are grouped and color-coded by their recommended time-of-action into three groups:

- **Before** – Indicates switch actions that should have been implemented in the past according to a previous recommendation of the system. These switch actions are kept for reference only and address the situation where the plant operator has not been able to implement the recommended action in time.

- **Current** – Indicates switch actions that are to be implemented in the current cycle time interval.

- **Future** – Switch actions whose time-of-action lies in the future. As the EWave system periodically updates switch recommendations, these entries are subject to change when future calculation cycles are completed.

The **values tab** displays detailed values for the variable that is currently selected in the component panel. Variables that are scalar-, vector- or matrix-valued are displayed as simple tables, while time series are plotted as charts.

The type of chart depends on the type of the underlying variable: A Boolean variable indicates an on/off switch profile and hence displayed as a bar chart, where periods of value *true* are shown as a solid bar and periods of value *false* as a transparent bar (see Figure 10.12). For all-time series where the underlying variable is of numeric type, a line chart is displayed (see Figure 10.13).

Figure 10.12. Values tab showing bar chart

Figure 10.13. Values tab showing line chart

All three panels of the EWave UI are functionally linked in such a way that

- If the user navigates to a page on the plant structure view, the component tree automatically jumps to that page and expands its components (and vice versa).
- If a component is selected in the plant structure view, it is highlighted and the component tree automatically jumps to selected component and expands its variables (and vice versa).
- If a variable is selected in the component panel, the details panel switches to the property tab and shows the value(s) of the selected variable.

In addition, whenever a cyclic run of the EWave DSS completes, the results panel switches automatically to the message tab and displays the recommended switch actions from this run. This ensures that the user always gets to see the results from the latest cyclic run.

10.2.4 Communication

The EWave user interface consists of a frontend application and a backend web server. The web server runs as an independent process that connects to the EWave execution system and retrieves updated calculation results as they are available.

The frontend is implemented as a *single page application* (SPA) running inside a web browser. All static UI resources such as page templates, layouts or graphics are retrieved from the web server using HTTP requests during the initial loading of the frontend in the browser. Subsequent page content is rendered dynamically during application runtime based on data retrieved on demand from the web server process.

For this dynamic loading of data, two mechanisms exist in the EWave system: Data loaded due to an active user request (e.g. by selecting a particular component or choosing a profile for viewing) is retrieved via AJAX-style requests triggered by a user interaction with some UI control. Furthermore, there is a push-mechanism that allows the EWave execution system to actively push data to the web server and subsequently to the frontend whenever a calculation is finished and new results are available.

To this end, a special type of calculation module, called UI module, is attached to the execution system that generates switch action messages, extract the part of the calculation results that are relevant to the user and handles persistent connections with the web server. The architecture of this mechanism is shown in Figure 10.14. The persistent connection between UI module and the web server is made through TCP.

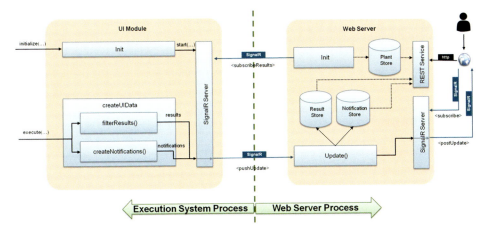

Figure 10.14. Architecture of the UI module push mechanism

Whenever a cyclic run of the EWave system is finished, the *execute()* method of the UI module is called, which generates the switch notifications and extracts the relevant result data before transmitting it to the web server, which persists the data in memory and notifies all connected web browsers of the updated results.

10.2.5 Technologies

The frontend of the EWave user interface consists of HTML-templates and some CSS stylesheets describing the layout of the website. The Bootstrap framework ([3]) has been used to simplify dealing with the layout for different resolutions and browser versions.

All functionality of the frontend has been implemented in Javascript and follows a *model view controller* (MVC) approach using the AngularJS library ([1]).

In the plant structure panel, navigation sheets and process view sheets are dynamically loaded as SVG elements directly into the browser's DOM, which allows defining handlers for user interaction (such as mouse clicks) with individual components. For drag and zoom functionality, the SVG-pan-zoom library ([10]) has been used. Sheets with a geoview are rendered using the OpenLayers library ([7]). The geoview is backed by a streetmap for which static maptiles have been obtained from the OpenStreetView project ([8]).

For implementing the component panel, the *TreeView*-component of the Kendo UI framework ([9]) has been used. Finally, the charts displayed in the values tab of the details panel are rendered using the Highcharts library ([4]).

The backend implementation of the user interface has been written in C# and is backed by the ASP.NET framework ([2]). Dynamic frontend content is delivered in JSON-format via HTTP requests. For this purpose, a REST API to the data model described in Section 9.2.3 has been implemented. Persistent connections to the UI module are based on the SignalR library ([9]), which has also been utilized to provide push notifications for updated switch action messages and new result data to the connected browsers.

Bibliography

[1] *AngularJS*, https://angularjs.org.

[2] *ASP.NET*, https://www.asp.net.

[3] *Bootstrap*, http://getbootstrap.com.

[4] *Highcharts*, https://www.highcharts.com.

[5] *Kendo UI*, https://www.telerik.com/kendo-ui.

[6] M. Plath and K. Wichmann, Energieverbrauch der deutschen Wasserversorgung, *DVGW energie | wasser-praxis* 7+8 (2009), 54–55.

[7] *OpenLayers*, http://openlayers.org.

[8] *OpenStreetMap*, https://www.openstreetmap.org.

[9] *SignalR*, http://signalr.net.

[10] *Svg-pan-zoom*, https://github.com/ariutta/svg-pan-zoom.

[11] *VGW-Information Wasser Nr. 77*, SVGW Handbuch Energieeffizienz.

Part IV

Outlook

Nowadays, water supply companies operate their supply systems via automatic systems as well as central control centers. The new decision support system calculates pump schedules by adding further information. The aim of this system is an energy- and cost-efficient operation while maintaining a high security of supply. The system has a graphical user interface to visualize schedules, energy costs, etc. A corresponding system documentation as well as an energy and operational data management are the basis for the underlying water supply model of the new energy management system. In terms of the supply system considered here, optimization yielded an energy savings potential of approximately 10%, based on historical data. Once the energy management system has been successfully implemented at RWW and the legal requirements have been clarified, the industry partners together with the research institutions providing scientific assistance, intend to develop a joint go-to-market strategy which will also allow small and medium-sized water suppliers to use the management tool which has been developed.

Chapter 11
Field test

Annelie Sohr, Constantin Blanck, Stefan Fischer, Michael Plath,
Moritz Allmaras, Tim Schenk, and Andreas Pirsing

Abstract. The validation of the EWave system has been performed in two different ways. Main focus was the installation as pilot application directly on site at the RWW facility in the waterworks Dorsten-Holsterhausen. Before the installation and usage various IT security requirements had to be fulfilled and the control center operators were introduced to the goals and the potential of the EWave project. Then the EWave system was initially run in parallel to the real operation, without applying the setpoints on the real plant. Finally in the last stage of this test proposed setpoints were applied to the plant control. In parallel to this pilot test a further validation has been performed with the focus to quantify the benefit of EWave. To be able to assess the EWave benefit as closely as possible a concept has been developed that compared simulation results of a real historic plant operation with a theoretical EWave optimized plant operation. Finally, the results of the whole EWave project and the two validation approaches are evaluated and summarized and the potential of the approach is assessed in an outlook.

11.1 Implementation and pilot application

Apart from the actual pilot application, using the knowledge gained, the test of the drinking water demand forecast was also part of the implementation and the pilot application.

The first test phase was the foundation for test phase II. In the first test phase, the main focus was on the program itself, as the phase served to get to know the program in detail and to identify the initial problems and errors. Furthermore, the switching recommendations have already been evaluated and checked for usefulness. Minor bugs that could be observed during this period were corrected during this phase and the boundary conditions were adjusted. Test phase I was also used to recruit the operating personnel for the implementation of the second test phase. This is where dates with all five shift allocations of the RWW were accomplished and the EWave assistance system was presented.

In test phase II, the switching recommendations proposed by the assistance system were discussed with the operating personnel and subsequently, if there were no concerns, applied. This was why the transparency created with respect to the operating personnel in test phase I was very important. It was only thanks to the good cooperation that the very good results were achieved in the second test phase.

11.1.1 Preparation of the pilot application

The project partners carried out a preliminary test phase including the validation of the assistance system operation on the PC provided by RWW. Thus, in March 2017, a PC with executable software was handed over to RWW.

Integrating such a PC into the RWW server world was difficult due to the strict requirements, especially the new IT security law [1]. All requirements from an IT security perspective were summarized in an eleven-page document. Before integrating the PC into the RWW-PDV network (process data processing), the computer had to be hardened. This included virus and malware scanning, uninstalling the redundant software and closing all communication ports. Subsequently, new firewall rules for remote desktop access from the RWW office network were defined. The EWave computer was installed in the RWW server room in the Mühlheim-Styrum-Ost waterworks. There it was physically connected to the RWW-PDV network and integrated into the DMZ (demilitarized zone) of the ACRON server (no communication with the control system network possible). This PC was provided with its own fixed IP address. RWW employees have remote desktop access to the computer and thus to the assistance system. The schematic flow chart is shown in the communication diagram Figure 11.1.

In principle, no direct access to the computer was possible via remote access. This was not necessary for the test operation, but would have been helpful for the installation of updates by the project partners.

The effort in the event of errors and thus the error analysis was therefore significantly higher. After an error that occurred, the project partners were first sent the error file (ErrorLog). Updates were then exchanged via a file server and installed by RWW itself.

The assistance system constantly required a current data extract. This included measurement and counting values of a total of 46 data points as minute values for the last 24 hours. This data extraction was automatically realized after installation in the server room via an ACRON export.

11.1.2 Test phase I

The first test phase started by connecting the assistance system to RWW's control technology environment in the server room. From this point on, the EWave assistance system was continuously supplied with current data for the calculation via the automatic export. This made it possible to perform and display a new calculation of the optimum switching actions every half hour. Since the switching recommendations were not applied during this test phase, the focus was on evaluating the calculated recommendations. In this context, attention was paid to the average number of recommendations displayed after each calculation step, the extent of the changes resulting

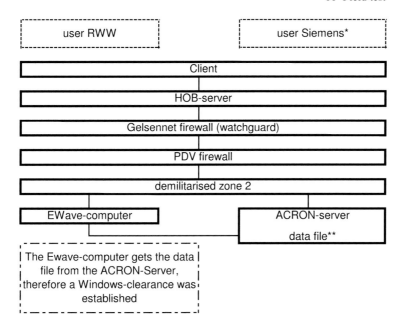

Figure 11.1. Communication diagram

from the switching recommendations and the calculated consequences resulting from the switching actions.

On average, the assistance system proposed nine switching recommendations. The general recommendations are listed in Figure 11.2. In addition, there were about six switching recommendations which could not be applied. The six switching recommendations were the valve positions of drinking water pumps not in operation and were therefore not relevant switching recommendations.

As an example, Figure 11.3 shows the adjustments of the volume flow for water treatment required by EWave for four successive calculations. The real volume flow over time is shown in black. The colored lines show the course of the volume flow calculated by EWave, based on the respective drinking water demand forecast belonging to the calculation. The area under each curve has to be the same so that in the end the same amount of water was treated.

The curves of the volume flow curves calculated by EWave are almost identical. For the calculations at 8:00 and 8:30, the assistance system suggested an increase of the volume flow to approx. $3,900\,m^3/h$ in each case. The increase should be made by 9:30. Subsequently, the volume flow would have been reduced to 2,800

Component	Effect
Galerie_HOL	Holsterhausen well flow rate
Galerie_UEF	Üfter Mark well flow rate
Ventil_UEF	Mixing ratio Üfter Mark
Ventil5_1	Processing quantity, per filter line
Ventil5_2	
Einschalten RWP X	Switching instructions of RWP
Ausschalten RWP Y	
Einschalten TWP X	Switching instruction of TWP
Ausschalten TWP Y	

Figure 11.2. Applicable switching recommendations

m^3/h by 14:30. According to the system, the volume flow should be increased again to approx. 3,900 m^3/h between 14:30 and 17:15. For the calculation at 9:30 and 10:00, the assistance system initially suggested a reduction of the volume flow to approx. 2,300 m^3/h. EWave proposed an increase of the volume flow to 2,800 m^3/h after approx. every 1.5 h. Comparable with the results of the calculations at 8:00 and 8:30, the volume flow should be increased to 3,900 m^3/h in the time window from 16:00 to 18:30 (calculation 9:30) or from 16:30 to 19:15 (calculation 10:00). The reason for the sharp increase and decrease in treatment resulted from the efforts of the optimization algorithm to fill the drinking water tank in an energy-efficient manner for the evening peak.

Figure 11.3 shows that the switching recommendations were constant and did not differ greatly from calculation step to calculation step. It has become clear here, however, that boundary conditions that limit the size of the change in the volume flow of water treatment are necessary. If the amount of water treated increases too quickly, the volumetric flow rates in the multi-layer filters change considerably. This results in a pulse that may loosen retained particles and shift them into the filter. After several pulses, this can lead to a breakthrough of particles. Besides the quality risk, which is increased by this, the filter service life is reduced and more frequent flushing becomes necessary. Another boundary condition that would limit the change in the treatment quantity is the minimum running time of the raw water pumps. In order to avoid signs of wear, care should be taken here to avoid frequent, unnecessary switching on and off of the units. Although it is not possible to quantify how much the unit is loaded by a switching operation, it has been observed in the past that frequently switched pumps have a greatly shortened service life. Boundary conditions such as maximum changes between two calculation steps and minimum runtimes as well as maximum runtimes, e.g. of pumps, can generally be stored in EWave. In this first test phase, however, this was dispensed with in order to keep the optimization model simple at first. Although the boundary conditions have a great influence on the operation, it was also interesting for the test to see what results the optimization delivers without

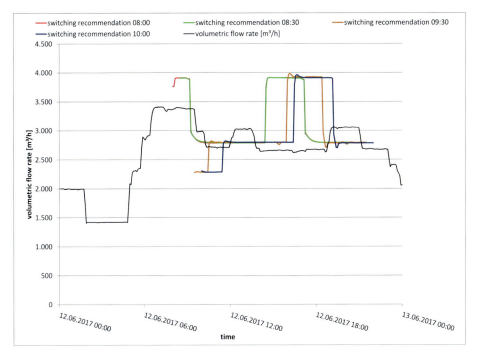

Figure 11.3. Representation of the switching recommendations of EWave, shown as an example for the processing volumetric flow rate

these boundary conditions. In some cases, the boundary conditions that were not stored were violated, but in others the boundary conditions would not have led to any limitation of the optimization.

Test phase I also served to analyze the behavior of the water reservoirs in detail before the entire system was to be switched in test phase II according to the EWave system's proposals. Figure 11.4 shows the water level of drinking water tank 1 at the Dorsten-Holsterhausen waterworks for the next twelve hours from the time of calculation. The water level in the drinking water tank is kept constant for four hours. Only when the drinking water output is increased in the evening hours from approx. 17:30, the drinking water tank is used successively to compensate for the peak demand.

Analogous to the behavior of the drinking water tanks, a positive summary could also be drawn from the analyses of the Üfter Mark tank. This storage tank serves as a buffer tank in order to be able to operate the extraction at a constant level even with increased raw water requirements. The wells of the Üfter Mark water extraction may not be switched as often due to the raw water quality. In addition, the mixing ratio of the Holsterhausen raw water and the Üfter Mark raw water of 60% to 40% must be maintained. Due to these two restrictions, practically only the Holsterhausen well gallery can increase the amount of raw water obtained, the share of the Üfter Mark well must be supplied from the raw water storage tank. In test phase 1, this

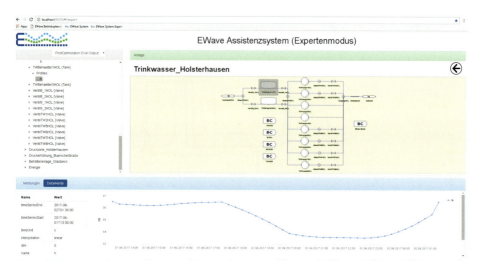

Figure 11.4. Depiction of the predicted water level in the drinking water tank 1 (screenshot)

management of the raw water storage tank worked similarly well as the management of the drinking water tank.

One of the most important processes in this first phase was to create acceptance of the system among control center staff. To this end, the different shifts were informed in personal briefings about the aim and the possibilities of the program. It was important not only to clarify the potential, but above all to take away the fear of substituting one's own job. Only in this way was it possible to convince the control center operators to contribute their extensive experience in handling the operation of the systems to the further development of the program in the second phase.

11.1.3 Test phase II

In the second test phase, the control station was equipped with an additional computer from which access to the EWave computer was possible. Additional personnel were deployed for operation, which operated the interface between the control station personnel and the assistance system.

Originally the support of early and late shift was planned, but this could not be fully implemented due to personnel reasons. The early shift was therefore supplemented by another employee one hour after the start from 7:00, and the late shift was accompanied until 8:00, i.e. until two hours before the shift change. After the shift change, the main operators change the systems to their personal preferences. These switching operations can partially influence the further course of the operation during the day.

Using another employee familiar with the EWave project to support the project made sense, since the internal naming in the EWave system first had to be translated into the names of the RWW control system. For example, the designation "Ventil1UEF" in the assistance system stands for the internal RWW designation "Förderung Zone Üftermark Aufbern-Regelarma. Öffnung". The additional employee familiar with EWave was also able to try to explain the purpose to the control desk personnel when EWave's switching requirements were not directly understandable. In some cases, however, the requirement could be implemented for test purposes. The vast professional experience of the control center personnel was largely in competition with the system. This is to be evaluated as very positive, since this made it possible to examine all recommendations etc. in depth. Operational safety and water quality are the priority of the control center operators' decisions, switches to increase energy efficiency are less frequently part of the considerations. Switching operations were therefore not implemented in some cases, as they were considered to be unfeasible and not to be useful. Later, these assessments of the control center staff could often be confirmed as correct. It was also only during the discussion about the possible purpose of the proposed switching action that the lack of some important boundary conditions became apparent.

The result of successful switching is shown in Figure 11.5. First, the drinking water pump 7 was replaced by drinking water pump 8 in order to react to the lower flow rate. As a result, the overall efficiency decreases in the short term, but then stabilizes at a slightly higher level. Shortly before 20:00, after a suggestion of the assistance system, the drinking water pump 2 was switched on in order to relieve pumps 4 and 7. Efficiency rose in the single-digit percentage range until the flow rate declined again.

Figure 11.5. Realization of the suggested switching operations

One possible cause for the, in some instances, proposed strong changes in quantities could be the long period between data provision and implementation of the switching recommendation. If current measured values from the control system were stored on the long-term archiving system via the coupling server, they were already up to 5 minutes old. The optimization took about 20 minutes, after which the results were displayed. Subsequently, the proposals of the assistance system were usually still explained or discussed. There were therefore various time delays between picking-up the exact, measured plant status and the implementation of the optimized control recommendations based on it. Although EWave tries to keep the plant status as up-to-date as possible by simulation-based extrapolation (see Chapter 9, Section 3.2), these effects need to be investigated even more closely. Active control of the water supply was therefore not yet sufficiently efficient under these conditions. However, this is not the primary goal for such a test.

11.1.4 Drinking water demand forecast test

The forecast is based on the drinking water output for the years 2013, 2014, 2015 and partly from 2016, which was provided by RWW in a time resolution of 15 minutes. On the basis of this data, a progress profile was created for each day and each hour of the year. The change from a workday to a public holiday, for example, was taken into account in these profiles. The forecast showed that, as expected, the weather and the climate also have a major influence on the course of drinking water output. However, since the weather is only forecast for the next 24 hours, a weather-independent calculation method was developed. The start value for the drinking water demand forecast is the average value of the drinking water output (minute value) of the last 60 minutes. This value can be used as an indirect indication of weather conditions. If the value is above average at the defined time, a higher drinking water output can be expected from a very warm day. Another advantage of this initial value is that it corrects forecast errors automatically. Because the forecast refers to real values throughout, the differences between forecast output values and real output values are very small. Figure 11.6 shows an example of the real drinking water output and the predicted drinking water output process at defined times for one day. Since the start value is based on the real values, a slight correction is made along the gradient line for each calculation loop.

11.1.5 Conclusion: pilot application

The pilot application showed that the assistance system can increase the energy efficiency during operation of the waterworks by means of forward-looking switching recommendations. Improvements would have to be made to the described time delay between collection of the measurement data and implementation of the switching recommendation, as well as to the comprehensive project planning of all necessary boundary conditions. With regard to the prototype character of the program, the test can be rated as successful under real conditions.

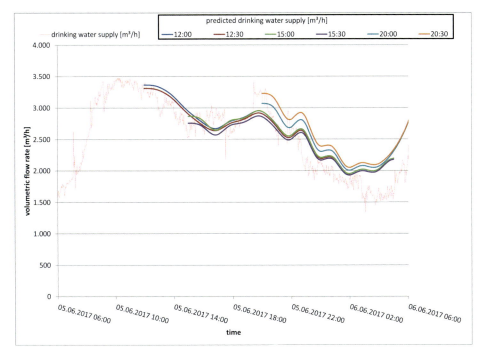

Figure 11.6. Comparison of the actual drinking water supply with the predicted drinking water supply

11.2 Comparative calculations

11.2.1 Concept

The previous chapter focused on the piloting of the EWave system at the real control center of the RWW to validate if the setpoint proposals of EWave are reasonable and qualitatively beneficial in terms of its defined goals. As the real plant can only be driven with or without EWave there is no possibility to quantitatively assess the energy savings EWave would provide. The only correct solution to that problem would be to have two identical plants running in parallel and one is controlled by the operators as of today and the other is controlled by EWave. The theoretical potential of the optimization has been outlined in Chapter 5. This chapter will highlight in more detail what aspects have to be considered when assessing the real potential of EWave and how this can be achieved.

Obviously, the assessment of the EWave potential can only base upon past days of operation and the recorded operational measurements of those days. The idea is to compare the real operation in those days with a fictitious EWave driven operation of those same days. EWave, unfortunately, does not evaluate an identical match of a real plant but uses several models and assumptions in its evaluation. There are several models of the physical behavior of the plant in the simulation and optimization

	Real operation	**Virtual models**
Behavior of plant	Real plant	Simulation and optimization models
Water demand	Measured demands	Demand prognosis
Initial state of the plant for optimization	Measured states	Extrapolation from most actual measurements (s. chapter 9.3.2)

Table 11.1. Different EWave models for certain relevant aspects of reality

modules, a model for the demand prognosis and an extrapolation of the plant behavior for the time interval between the most actual measurements and the initial time point of the optimized setpoints (see Table 11.1).

Comparing a past real plant behavior with a fictitious EWave evaluation for the same past period should avoid including modeling errors as far as possible or at least compare only values that include the same modeling errors otherwise the results cannot be interpreted anymore.

First of all, it is clear that the effect of the optimized setpoints for past operation time periods can only be simulated. This means a comparison of real plant values for energy consumption with simulation-based values would also include the comparison of the simulation model with real behavior. To avoid this, the real plant values are replaced by a simulation using the real operating setpoints for that time period. This real plant simulation has to be as close to the real behavior as necessary for a correct comparative evaluation. In EWave there is the possibility to adapt to the real measured filling levels as it is done during the state calculation (s. Section 9.3.2). This kind of simulation is now utilized for the simulation of the real plant behavior for the whole optimization horizon as well so that the modelling errors are avoided because of which water is generated or lost in accordance to the real plant behavior.

The second step is to calculate the behavior of the whole plant under optimized setpoints. This process follows the same procedure as EWave and would perform under real operation conditions, except for the following aspects:

- The optimization requires minimum tank filling levels at the end of the optimization horizon. These now have to be set to the measured values at this time point to ensure that the whole water quantity that is moved in the optimization period is the same as it was in the real plant.
- The initial state for the optimization is not extrapolated as in Section 9.3.2, but can be simulated exactly by the simulation state calculation utilizing the measurements of filling levels, operational setpoints and demands.
- The optimization is performed twice. One is carried out with the water demands calculated by the demand prognosis. The other calculates the optimal setpoints using the measured demands.
- The simulation applying the optimized setpoints is then also carried out twice with the two different produced sets of control setpoints, but the simulation always uses the measured demands and not the demand prognosis.

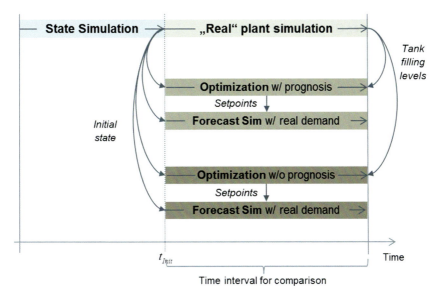

Figure 11.7. Generation of comparable plant behavior

An overview of the full process is shown in Figure 11.7.

This approach guarantees that the values to compare include as less modeling errors as possible by using the realized EWave modules. Possible evaluations are

- the difference in calculated optimized setpoints, on the one hand using real measured data for the water demand and on the other hand the water demand prognosis,
- the potential of a single EWave evaluation by comparing the simulation of the real plant operation and the simulation including optimized setpoints derived by measured water demands,
- a comparison closer to the real EWave application by comparing for a single EWave evaluation the simulation of the real plant operation and the simulation including optimized setpoints derived by predicted water demands.

11.2.2 Verification

This comparison has been executed on a historical data set from March 14th, 2017. The considered time horizon is 24 hours starting from 6:00 in the morning and thus lasting till 6:00 March 15th, 2017. It has to be noted that the energy price was assumed to be fixed for the whole scenario and that only a few boundaries for pump switching times have been set. Thus the potential the optimization could exploit was not much. Still the results should be presented here to show in general the validity of this comparison approach.

The total water demand at the outlet of the waterworks Dorsten-Holsterhausen for that day was measured to be $16.8\,\mathrm{m}^3$, the demand prognosis predicted an around

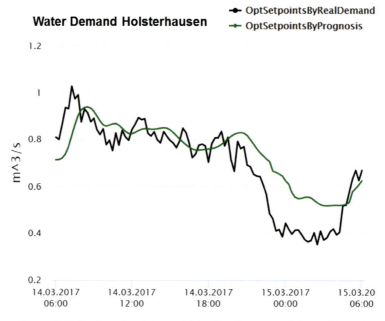

Figure 11.8. Profiles of real demand (black) and prognosed demand (green)

seven percent higher water demand of 18 m³. The distribution of the water demand that day is shown for both time series in Figure 11.8.

The total energy demand of that day evaluated by the real plant simulation was 37.711 kWh. Despite assuming more water demand the optimization basing on the demand prognosis resulted in around the same energy consumption with 37.594 kWh. An optimization with the real water demand would result in around four percent less energy demand, namely 36.344 kWh that day. In Figure 11.9 it is indicated by the red arrows that the optimization mainly switches probably more ineffective pumps off and uses others. As we do not have a dynamic energy price in this scenario, this is the main achievement a single optimization run with a 24-hour horizon can accomplish.

Figure 11.10 displays the filling level of one of the drinking water tanks in waterworks Dorsten-Holsterhausen and points out very clearly the strategy of the single optimization run. The tanks are held at a minimum required level as long as it suites the goal to minimize the energy. At the end of the horizon interval, the optimization finds the suitable spots to pump water into the tanks to fulfill the water demand peak for the next morning. As the demand prognosis estimates a higher water demand than there really was, the inflow to the tanks is accordingly higher.

This effect would be smoothened out if this comparative study would take the full EWave approach into account, namely that EWave is called cyclic and produces probably every 30 minutes new setpoints adapted to the actual plant state and a new water demand prognosis. For such a comparison a concatenation of all EWave cycles, which includes the time interval of each EWave cycle till the next EWave cycle starts, is necessary and would then represent the EWave controlled plant behavior over the

name of comp / total	energy [kwh] Real	energy [kwh] OptSetProg	energy [kwh] OptSetReal
EnergiebedarfGalerieUEF	3473.5	2718.3	3060.8
EnergierueckgewinnungUEF	-1102.4	-651.9	-856.2
EnergiebedarfGalerieHOL	6601	7609.3	6596.9
EnergiebedarfRiesler1HOL	132.5	132.5	132.5
EnergiebedarfRiesler2HOL	132.5	132.5	132.5
EnergiebedarfUV1HOL	407.5	377.5	375
EnergiebedarfUV2HOL	418.8	377.5	375
RWPumpe1HOL	969	1188.8	1177.2
RWPumpe2HOL	0	1010.8	1056.9
RWPumpe3HOL	1289.3	1630.8	1457.8
RWPumpe4HOL	0	0	0
RWPumpe5HOL	2287.6	51	77
RWPumpe6HOL	0	0	0
TWPumpe1HOL	5346.2	1420.2	186
TWPumpe2HOL	0	1064.2	1149.8
TWPumpe3HOL	0	0	0
TWPumpe4HOL	10552.5	10598.8	12531.8
TWPumpe5HOL	0	0	0
TWPumpe6HOL	0	0	0
TWPumpe7HOL	6566.1	283.1	296.6
TWPumpe8HOL	0	9467	8413.9
DEA1BuS	22.8	77.8	72.7
DEA2BuS	121	106.6	108.4
Pumpe2GLA	0	0	0
Pumpe3GLA	42.4	0	0
Pumpe1GLA	451.1	0	0
Pumpe4GLA	0	0	0
total	37711.4	37594.8	36344.6

Figure 11.9. Detailed overview of the simulated energy consumption of the scenario

whole time horizon Figure 11.11. But the potential of this approach can only be measured if the optimized setpoints are really applied on the plant. This means the plant has to be operated with the EWave setpoints till the time point, where the next EWave cycle starts, and then the actually reached plant state at this time point has to be considered for the next EWave cycle. As there is no real plant to apply the optimized

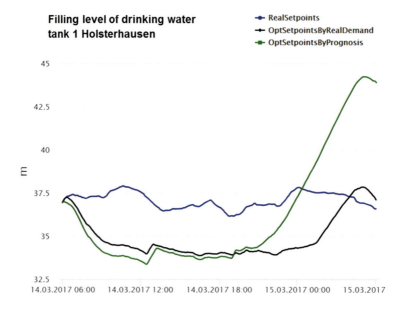

Figure 11.10. Filling level profile of the tank 1 in Dorsten-Holsterhausen

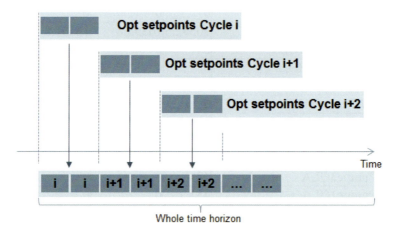

Figure 11.11. Concatenated EWave setpoint proposals

setpoints and retrieve plant measurements for the next EWave cycle this initial state for the next EWave cycle has to be calculated by a simulated plant behavior without any adaption to real filling levels and thus includes modeling errors in the initial plant states.

So the conclusion is, that all those comparison approaches are quite useful to get at least some assessment of the benefit of the EWave system. But as there are no two identical plants available that could be driven with and without EWave in parallel, there are always models in use, which include assumptions and errors. There is a much cited aphorism of George Box about statistical models that hits the point very well: *Essentially, all models are wrong, but some are useful* [1].

11.3 Conclusion & outlook

The operation of water supply systems is complex. On the one hand, the water supply system company must ensure that consumers are always adequately supplied with drinking water whereas on the other hand, consumers demand adequate water prices in addition to security of supply.

The assistance system developed within the framework of the project EWave enables water supply system companies to calculate cost-optimal operating modes for the controllable aggregates with the help of mathematical simulation and optimization methods. Important boundary conditions for optimized operation are knowledge of current and future drinking water and energy requirements. The energy requirements depend on the hydraulic conditions in the waterworks and on the considered distribution network. Therefore, a drinking water forecasting model and a fully dynamic hydraulic simulation model were developed.

The hydraulic simulation model is based on a network approach in which the network elements such as pipes, reservoirs, valves, pumps, etc. are described by mathematical equations and coupled with each other via a network graph. The resulting nonlinear, differential-algebraic equation system is automatically generated and solved numerically using suitable methods. To account for the entire drinking water network in the dynamic simulation model, a network abstraction is necessary. This is done by regionally combining subnetworks to superordinate pipe and storage elements, thereby preserving the large-scale network structure.

EWave uses a newly developed, integrated optimization module for decision and operational support. As a result, the user receives plant timetables in a 30-minute time frame. For this purpose, mathematical optimization methods are used, which combine discrete-linear and continuous-nonlinear methods. First, a mixed-integer optimization model is solved in order to derive all discrete decisions (primarily pump schematics). EWave uses the results for the discrete optimization quantities and subsequently optimizes the continuous variables such as pump speed, valve opening degrees, or water quantities. Energy-optimized operation requires a cross-component, global view of the entire system.

The mathematical optimization methods used in EWave are not dependent on already known pump and plant schedules. Rather, it turns out that in this way other pump schedules and beyond that, control variables can be proposed, which lead to a more energy-efficient overall operation. Offline optimization results yield an energy saving potential of up to 10% for the pilot plant's waterworks.

In practical testing at RWW, it was possible to reduce the energy consumption by approx. 5% in the considered subnetwork. Due to the fact that RWW currently does not use dynamic electricity prices, the optimization methods used cannot fully exploit their performance. The achieved reduction of the energy purchase is achieved exclusively by an optimal selection of the pumps. Greater cost savings are to be expected if the optimization process is additionally given the opportunity to use dynamic electricity prices.

Another advantage for the water supply system companies, which is difficult to assess financially, is the fact that EWave calculates cost-optimal results within a short time, even in the case of unusual and / or previously unobserved boundary conditions: electricity costs are optimized for normal operation. This takes account of high and low tariff periods in the context of electricity supply contracts as well as the power limits agreed with the energy supply company (kWmax monitoring). Here, the concerns of operating cost reduction with absolute security of supply are in the foreground.

The emergency strategies are aimed primarily at security of supply, since here it is even more important than in normal operation for the supplier that water can be provided in sufficient quantity: With the optimization calculations, it is possible to simulate and assess various accident scenarios, e.g. failures of individual pumps in pumping stations, pipe breaks in main lines or unexpectedly high water withdrawals at individual points in the network. This results in various emergency strategies that can be evaluated. On this basis, the available safety reserves can be used better. And the appropriate response of plant personnel to incidents can mitigate extremely costly emergency response.

EWave requires significantly less computation time compared to previous systems. Thus, the calculation of a current timetable on a personal computer with Pentium processor takes a maximum of 20 min. This makes it possible to initiate the optimization calculation automatically at short intervals and / or event-dependent and to constantly present a current proposal (suggestion) to the plant personnel.

Bibliography

[1] G. E. P. Box and N. R. Draper, *Empirical Model-Building and Response Surfaces*, John Wiley, 1987, 424.

Acronyms

ADM	alternating direction method
AI	artificial intelligence
API	application programming interface
BC	boundary condition
BMBF	Bundesministeriums für Bildung und Forschung (German Federal Ministry of Education and Research)
CPS	cyber physical system
CPPS	cyber-physical production system
DAE	differential-algebraic equation
DMZ	demilitarized zone
DOM	document object model
DSS	decision support system
EMS	energy management system
ERP	enterprise resource planning system
ENTSO-E	European Network of Transmission System Operators for Electricity
EWave	energy management system water supply
GIS	geographic information system
GUI	graphical user interface
ICT	information and communication technology
ID	identification
IOT	internet of things
IT	information technology
KPI	key performance indicator
KRITIS	Kritische Infrastrukturen (critical infrastructures)
LAN	local area network
LOD	level of detail
MES	manufacturing execution system
MILP	see MIP
MINLP	mixed-integer nonlinear program
MIP	mixed-integer (linear) program

ODE	ordenary differential equation
OT	operational technology
PADM	penalty alternating direction method
PDAE	coupled system of PDE and DAE
PDE	partial differential equation
P&ID	process and instrumentation diagram
PLC	programmable logic controller
REST	representational state transfer
RMS	root mean square
RWW	Rheinisch-Westfälische Wasserwerksgesellschaft mbH
SCADA	Supervisory Control and Data Acquisition
SPA	single page application
SVG	scalable vector graphics
UV	ultra violet
VPN	virtual private network
XML	extensible markup language

Symbols and Parameters

symbol	meaning	unit
A	cross-sectional area (or set of edges)	m² (-)
D	hydraulic diameter	m
g	gravitational acceleration	m/s²
H, h	pressure head	m
H_L, h_L	pressure head at inlet (left end)	m
H_R, h_R	pressure head at outlet (right end)	m
H_g	groundwater level	m
k	pipe roughness	mm
L	length (of pipe/element)	m
n	revolution speed	1/s
P	power (consumption)	kW
P_{el}	electric power (consumption)	kW
P_{hyd}	hydraulic power (consumption)	kW
Q, q	volume flow	m³/h
Q_L, q_L	volume flow at inlet (left end)	m³/h
Q_R, q_R	volume flow at outlet (right end)	m³/h
Re	Reynolds number	1
s	opening degree in [0, 1] (of valves) or filling level/height (of tanks)	1 or m
t	time, duration, time span	h
v	velocity	m/s
V	volume (or set of nodes)	m³ (–)
α	generic curve coefficients	context dependent
β	generic curve coefficients	context dependent
ζ	pressure loss coefficient	1
η	efficiency	1
λ	friction coefficient, friction factor	1
χ	characteristic function	–
ω	relative speed (of a pump) w.r.t. nominal speed	1

Glossary

Backwashing Removing of entrapped solids by reversing the flow of water through the filter media.

Chlorination Adding of chlorine to the water for disinfection.

Design pressure Highest operating pressure of the system or a pressure zone taken into account for calculations, but without consideration of pressure surges.

Disinfection Treatment process for decontamination of fluids and surfaces. There are several techniques available, such as ozone disinfection, UV disinfection and chlorination.

Filtration Separation of solids from liquid by using a porous substance that only lets the liquid pass through.

Freshwater Water containing less than 1 mg/l of dissolved solids of any type.

Groundwater Water that can be found in the saturated zone of the soil; a zone that consists merely of water.

Mainline Water pipe with main distribution function within one supply usually without a direct connection to the consumers.

Potable water Water that has passed the entire water treatment and is suitable for human consumption, according to national regulations.

Pressure zone Zones with different energy levels within a water supply system.

Pre-treatment Processes used to reduce pollutants from the raw water.

Product water Pure water which has passed through a water treatment plant, but still not ready to be delivered to consumers.

Pumping station Conveying system to ensure sufficient pressure and flow within the water distribution system. There are three types. Main pumping station: Usually after the treatment plant or, if no treatment is carried out after the recovery to ensure the water transport. Intermediate pumping station: to ensure the water transport to a reservoir or supply area; Booster station: to increase operating pressure within supply areas without storage.

Raw water Intake water which has not yet been processed.

Reservoir A natural or artificial storage area used to store water.

Service pressure Pressure at zero flow at the transfer point to the consumer.

Storage tank Closed storage system for drinking water which ensures pressure stability and compensates consumption fluctuations.

Surface water All water naturally open to the atmosphere, e.g. rivers, lakes, reservoirs, ponds, streams, seas, and wetlands.

Ultra-violet oxidation Disinfection process using extremely short wave-length light that can destroy micro-organisms (disinfection). Ultra-violet light has a wavelength shorter than visible light.

Water distribution Part of a water supply system with pipes, drinking water tanks, conveyors and other facilities to distribute water to consumers. The system starts after the water treatment plant or, if no treatment takes place, after the water intake and ends at the point of delivery to the consumer.

Water extraction Water extraction starts with the entry of the water into the water intake and ends with the transfer to the water treatment.

Water quality The condition of water with respect to the amount of impurities in it.

Water supply Supplying the necessary energy, which enables the water to overcome the resistances within the whole network, so that at all transfer points the required amount is available with sufficient pressure.

Water treatment The water treatment is defined as treatment of the water, in order to adapt its nature to the respective purpose of use and to specific requirements.

List of Contributors

Dr. Moritz Allmaras
Siemens AG, Otto-Hahn-Ring 6, 81739 Munich, Germany,
e-mail: moritz.allmaras@siemens.com

Constantin Blanck
RWW Rheinisch-Westfälische Wasserwerksgesellschaft mbH, Am Schloß Broich
1-3, 45479 Mülheim an der Ruhr, Germany, e-mail: constantin.blanck@rww.de

David Dreistadt
Hochschule Bonn-Rhein-Sieg, Grantham-Allee 20, 53757 Sankt Augustin, Germany,
e-mail: david.dreistadt@h-brs.de

Stefan Fischer
Netzgesellschaft Düsseldorf mbH, Höherweg 200, 40233 Düsseldorf, Germany,
e-mail: stfischer@netz-duesseldorf.de

Dr. Björn Geißler
Friedrich-Alexander-Universität Erlangen-Nürnberg, Cauerstraße 11, 91058
Erlangen, Germany, e-mail: bjoern.geissler@fau.de

Patrick Hausmann
Hochschule Bonn-Rhein-Sieg, Grantham-Allee 20, 53757 Sankt Augustin, Germany,
e-mail: patrick.hausmann@h-brs.de

Tim Jax
Hochschule Bonn-Rhein-Sieg, Grantham-Allee 20, 53757 Sankt Augustin, Germany,
e-mail: tim.jax@h-brs.de

Prof. Dr. Oliver Kolb
University of Mannheim, Department of Mathematics, D-68131 Mannheim,
Germany, e-mail: kolb@uni-mannheim.de

Prof. Dr. Jens Lang
Technische Universität Darmstadt, Dolivostr. 15, 64293 Darmstadt, Germany,
e-mail: lang@mathematik.tu-darmstadt.de

Prof. Dr. Alexander Martin
Friedrich-Alexander-Universität Erlangen-Nürnberg, Cauerstraße 11, 91058
Erlangen, Germany, e-mail: alexander.martin@fau.de

Dr. Antonio Morsi
Friedrich-Alexander-Universität Erlangen-Nürnberg, Cauerstraße 11, 91058
Erlangen, Germany, e-mail: antonio.morsi@fau.de

Dr. Andreas Pirsing
Siemens AG, Nonnendammallee 101, 13629 Berlin, Germany,
e-mail: andreas.pirsing@siemens.com

Dr. Michael Plath
RWW Rheinisch-Westfälische Wasserwerksgesellschaft mbH, Am Schloß Broich
1-3, 45479 Mülheim an der Ruhr, Germany, e-mail: michael.plath@rww.de

Roland Rosen
Siemens AG, Otto-Hahn-Ring 6, 81739 Munich, Germany,
e-mail: roland.rosen@siemens.com

Tim Schenk
Siemens AG, Otto-Hahn-Ring 6, 81739 Munich, Germany,
e-mail: tim.schenk@siemens.com

Dr. Annelie Sohr
Siemens AG, Otto-Hahn-Ring 6, 81739 Munich, Germany,
e-mail: annelie.sohr@siemens.com

Prof. Dr. Gerd Steinebach
University of Appplied Sciences Bonn-Rhein-Sieg, Grantham-Allee 20, D-53757
Sankt Augustin, Germany, e-mail: gerd.steinebach@h-brs.de

Lisa Wagner
Technische Universität Darmstadt, Dolivostr. 15, 64293 Darmstadt, Germany,
e-mail: wagner@mathematik.tu-darmstadt.de

Maximilian Walther
Friedrich-Alexander-Universität Erlangen-Nürnberg, Cauerstraße 11, 91058
Erlangen, Germany, e-mail: walther@math.fau.de

Index

active components, 95, 99
advanced manufacturing, 4
alternating direction method, 91
Anaconda, 52, 63, 75
architecture, 152
artificial sink, 108, 111, 120
artificial tank, 108, 111, 118
atypical grid usage, 22, 23

benchmark, 187
big data, 6
boundary conditions, 33, 60

calibration, 110, 114, 129, 133, 145
    network model, 137
    pump curve characteristics, 133
    valve coefficients, 135
carbon dioxide, 21
cold start, 61
combinatorial decision variables, 73
component, 156
component hierarchy, 157
component-based, 171
component-based modeling, 31, 51
connected components, 89
connection, 53, 87, 156
constraint
    coupling, 99
constraints
    long-term, 171
    nonconvex, 75
    nonlinear, 73, 75
    operational, 171
continuous optimization, 73
control energy, 22, 23
coupling condition, 58
cyber-physical production systems, 7
cyber-physical systems, 7
cyberattacks, 9
cyclic evaluation, 177

DAE solver, 62
data analytics, 6
data mining, 7
data provision, 147
decision support system, 11, 19
decomposition, 89, 100
demand forecast, 39, 208
demand rate, 22
deployment, 153
digital twin, 6, 29, 152
digitalization, 29
directed graph, 52, 76
discrete optimization, 73
Donlp2, 75
Dorsten-Holsterhausen, 25, 39, 99, 129, 148, 170, 186, 201

edge, 52, 78, 100, 157
efficiency
    electrical, 184, 187
    energy, 183
    hydraulic, 185, 187
    overall, 186, 187
electric power consumption, 57
electricity
    exchange, 22
    market, 20
    price forecast, 149
    procurement, 22
    stock exchange, 22
energy efficiency, 19
energy rate, 22
energy transition (Energiewende), 21
enterprise resource planning, 5
EWave, 19, 151
EWave-DOPT, 75, 99
EWave-NOPT, 75
EWave-OPT, 73

geographic information system, 193
graph, 52, 76, 157

Gurobi, 75, 99

Holsterhausen, 25, 99, 129
horizontal integration, 5
hydraulic model, 52
hydraulic modeling, 51

implementation, 201
Industry 4.0, 3, 21, 152
Industry 4.0 platform, 3
information and communication technology, 4
initial state, 32
initial values, 61
integrated engineering, 6
internet of things, 5, 8, 164
Ipopt, 75, 99
IT security, 8

junction, 78

key performance indicator, 183
Knitro, 75

Lax-Friedrichs approach, 60

Made in China 2025, 4
manufacturing execution systems, 5
mathematical optimization, 29
MATLAB, 45, 108, 135
microservice, 164
MINLP, 75, 89, 99
MIP, 75
MIP relaxation, 96, 99, 100
mixed-integer nonlinear programming, 75
mixed-integer programming, 75
mode
    expert mode, 190
    operator mode, 190
model-based systems engineering, 6

network aggregation, 108, 110
network simplification, 108
NLP, 75, 99
node, 52, 77, 100, 157

nonconvex, 91
nonlinear optimization, 73
nonlinear programming, 75

objective function, 88
operation planning, 174
operator support, 31
optimization, 29, 73, 167
ordinary differential equation, 53
OT systems, 8

parameter, 157
partial differential equation, 53
partial minimum, 91
passive components, 95, 99
penalty alternating direction method, 89
penalty function, 92, 173
piecewise linear approximation, 73, 96
pilot application, 201
pilot network, 25, 99, 129, 162
pipe, 54, 79
pipe aggregation, 116
pipe reduction, 109
pressure head, 52
pressure tank, 56
process and instrumentation diagram, 192
profile, 158
prognosis
    demand, 33, 175
    energy prizing, 33
    own-energy generation, 33
pump, 54
    control pump, 54
    pump curve, 133
    pump schedule, 32, 144

receding time horizon, 30
Remez' Algorithm, 97
REST API, 197
Rheinisch-Westfälische Wasserwerksgesellschaft, 19

security of supply, 19
single page application, 196
single source of truth, 152

specific energy, 183
state, 158
state variables, 52
subgraph, 89
sustainability, 183

tank, 55, 78
    pressure tank, 56
time horizon, 177
trading
    day-ahead, 23
    short-term intraday, 23
TWaveGen, 120, 129
TWaveProg, 144
TWaveSim, 52, 57

usability, 190
user interface, 190
user role, 189
    engineer, 189
    manager, 189
    operator, 189

valve, 53, 79
    backflow preventer, 53
    check valve, 81
    control valve, 53, 82
    gate valve, 80
vertical integration, 5
visualization, 159
volume flow, 52

warm start, 61
water hammer equations, 79
water production, 130
water quality, 19
water treatment, 130
WENO scheme, 59